ACUPUNCTURE

The Ancient Chinese Art
of Healing
and How It Works Scientifically

ACUPUNCTURE

*The Ancient Chinese Art
of Healing
and How It Works Scientifically*

Completely Revised Edition

FELIX MANN, M.B.

 VINTAGE BOOKS

A Division of Random House, New York

Library of Congress Cataloging in Publication Data
Mann, Felix.
  Acupuncture.
  1. Acupuncture.
[RM184.M34 1972]   615'.89   72-5837
ISBN 0-394-71727-9

CONTENTS

ACKNOWLEDGEMENTS

All inventions, all methods, are built upon the accumulated experience of others, of this and earlier generations.

For acupuncture I must thank most of all the Chinese of prehistoric and historic times, most of whose names are unknown to me. The early and modern Chinese books that I have used are mentioned in the various bibliographies. This book is largely based on the Zhongyixue Gailun of the Nanking Academy of Chinese Medicine, and the Zhenjiuxue Jiangyi of the Acupuncture Research Section of the Shanghai Academy of Chinese Medicine.

I myself first studied acupuncture under Dr Strobl of Munich and Dr Bischko of Vienna (President of the Austrian Acupuncture Society). Without them, particularly Dr Bischko, I might never have started; their skill, patience and generosity of ideas have given me a firm foundation. Later studies took me to Dr Van Nha and I have had many suggestions from Dr Manaka. Dr Schoch of Strasbourg, France, whose assistant I was for six months, taught me the art of diagnosis from patients' appearance, gait, colour, expression and atmosphere, in the tradition of Paracelsus.

I would also like to thank the Chinese Medical Association and the many doctors of Peking, Nanking and Shanghai for their hospitality whilst I was in China. Their untiring efforts in answering all my questions and in showing me their special techniques has been of inestimable value.

Before I knew Chinese, Charles Curwen did some translations for me. David Owen was my first teacher of the Chinese language, enabling me to read my Chinese acupuncture books. Later Frank Liu continued, and we meet every week, for Chinese is the type of language that is forgotten by a European if not continually practiced. Most of the Chinese has been Romanised using the Chinese national phonetic system. Latterly this has been used less in China. Therefore in those parts of the book which were written last, the sources of the quotations have been simplified by giving only the number of the volume and chapter. The few quotations which are retained from the first edition are mainly from the books of Veith, Soulié de Morant, Chamfrault and Hübotter, mentioned in the bibliography.

All European acupuncturists owe Soulié de Morant a debt for his original translations of Chinese treatises, and what is more, his understanding of the subject and its practical application in which he was associated with Dr Ferreyrolles. Before I was able to read Chinese I used and greatly benefited from the majority of French and German books on acupuncture mentioned in the bibliography.

I would like to thank Mr Alexander who granted me facilities to do acupuncture on suitable patients for four years in his Ear, Nose and Throat department at St. James' Hospital.

The writings of Rudolf Steiner enabled me to understand what the Chinese mean by their references to life-giving forces and the interplay of the material and non-material. Arthur Byers helped me on a much needed occasion, and also since then.

Mrs Francis Temple Thurston read the manuscript of the first (and parts of the second) edition of this book, and made many helpful suggestions and corrections. Frederic Metcalf made the first edition drawings. Mr Ronald Fuller, who has ghost-written some well-known books, took over the stupendous task of this second edition, which contains a lot of material directly translated from the Chinese, and has managed to put it into reasonable English without losing the original Chinese meaning. Alan Grant read through and corrected the manuscript of the 1st chapter. Sylvia Treadgold, the medical artist, has made the second edition drawings.

Most of all I would like to thank my patients who have had the courage to try something that is new in England, and who have had the equal courage to talk about it in the face of criticism. The response and attitude of a patient is something that a doctor, who looks upon his work as his vocation, appreciates above all else.

My mother has helped me more than any other person; for, if she had not taught, coached, and trained me in a way in which no one else could have done to see not only what is factual, but also what is 'beyond' the fact, I would never have developed the sensitivities that are required to practise acupuncture fully. Many new ideas, compassing a large vista that have an uncanny depth of perception and that both in general theory and in specific medical detail are true to life, have been hers, not mine.

Finally, I wish to thank the publishers especially Mr Owen R. Evans and Miss Ninetta Martyn.

The case histories mentioned in the text are from my practice. In most instances I have tried to select typical cases, to illustrate and enliven the theoretical part.

As yet scientific Western medicine has only penetrated certain aspects of traditional Chinese acupuncture. Hence the rather extensive parts of this book which are traditional have been largely taken from the Chinese sources mentioned in the bibliography. No European could use language of that type of his own accord.

My main contribution to acupuncture has been to find out how it works from a scientific point of view. This is described, for the first time in detail, in the first chapter of this book. Various minor ideas concerning needle technique, embryology, the invalidity of most categories of acupuncture points, meridians, laws of acupuncture etc. etc. have also been mine.

PREFACE TO SECOND EDITION

At the time when the first edition of this book was written, hardly anything on the subject had appeared in English. For this reason, although the book was primarily written from a medical point of view, the language used was such as to enable the layman to read it without difficulty.

A few months before the publication of the 2nd edition of this book, my more popularly written book on acupuncture for the layman, 'Acupuncture — Cure of Many Diseases' was published. Hence this second edition is rewritten more from the point of view of doctors and others having a greater interest in Far Eastern philosophy and medicine, as my other books (except *Acupuncture — Cure of Many Diseases*) have always been.

This edition contains about 50% new material. A few sections are completely deleted. Some sections of the 1st edition remain, although for the above mentioned reason, their vocabulary is less technical. A few parts, of an introductory nature, are taken from my popular book and hence demand less effort of the reader.

Although this book highlights the virtues of acupuncture and mentions some of the failures of orthodox Western medicine, let no one imagine that I underrate ordinary medicine. Indeed if my practice were in the country I would combine both schools, using whichever method was better in a particular patient or disease. Since I am, though, in central London, I tend to specialise in those diseases where acupuncture is the better method of treatment; combining the acupuncture (which will not be apparent from this book) with Western physiology, pathology, diagnostic methods and if appropriate ordinary drugs.

LONDON, W.I. 1971   FELIX MANN

ACUPUNCTURE

The Ancient Chinese Art
of Healing
and How It Works Scientifically

I

GENERAL CONSIDERATIONS

Acupuncture is an ancient Chinese system of medicine in the practise of which a fine needle pierces the skin to a depth of a few millimetres and is then withdrawn. The only thing of real importance in the study of acupuncture is to know at what point to pierce the skin in relation to which disease.

The notion that a pinprick, often in a part of the body far removed from the seat of the disease, can cure an illness is alien to conventional thinking. It is unfortunately the case that many doctors, even when faced with one or several patients who have been cured by acupuncture where their own efforts have been fruitless, refuse to believe the evidence.

The oldest records of acupuncture (acus = needle, punctura = puncture) are to be found on bone etchings of 1600 B.C. The first book of acupuncture, which contains a wealth of detail, is the Hungdi Neiging Suwen written about 200 B.C. It is one of the earliest treatises in Chinese on any subject.*

Acupuncture is not the exclusive possession of the Chinese. The papyrus Ebers of 1550 B.C. is the most important of the ancient Egyptian medical treatises. It refers to a book on the subject of *vessels* which could correspond to the 12 meridians of acupuncture. These vessels certainly could not refer to the arteries, veins or

*Hungdi Neiging Suwen—translated by Professor Ilza Veith as "The Yellow Emperors' Classic of Internal Disease". Published by Williams and Wilkins of Baltimore, 2nd edition.

nerves of the four limbs of the human body. However, my enquiries at the Egyptian Department of the British Museum have not been able to make matters clearer as the ancient Egyptian language is not well enough known to distinguish between the words for 'vessel' and 'meridian'.

The Bantu of South Africa sometimes scratch certain parts of the body to cure disease. In the treatment of sciatica some Arabs cauterise with a hot metal probe a part of the ear. This practice probably corresponds to a lesser known form of acupuncture called Ear acupuncture. Some Eskimos practise simple acupuncture with sharp stones. An isolated cannabalistic tribe in Brazil shoot tiny arrows with a blowpipe at specific parts of the body. The only observer ever to have returned from them thinks that, as the tribe show distinct Mongoloid features, this might also be related to acupuncture. Possibly the cautery practised in mediaeval Europe is also related to the tradition though this was mainly applied at congested or painful places and would therefore correspond to the simplest form of acupuncture in which only the *locus dolenti*, and not the distant part, is stimulated.

The great contribution of the Chinese to the primitive, or probably largely local form of acupuncture mentioned above, is that they have developed a fairly complete systematic method. Catalogued and described in numerous text books, it is taught at university and is reproducible at will under experimental conditions.

In China today the intending medical student can enter a university to learn, as in almost all parts of the world, Western medical practice. Or he can choose to study traditional Chinese medicine in another department of the same university. The student who chooses the Western path also studies the rudiments of the Chinese tradition while the other also follows courses in basic anatomy, physiology, pathology and other modern basic disciplines. Both courses take three to five or even seven years. Very few doctors of one school are also experts of the other, but in some hospitals doctors of each medical culture work together: the surgeon performs the operation, the acupuncturist treats the post-operative retention of urine, thus obviating the need for a catheter, and stimulates the lungs to prevent post-operative pneumonia.

Many Chinese are over impressed by Western medicine for they see that everything that has impelled China, or indeed any other non Western civilisation, into the twentieth century originated in the Western world. Quite literally, everything of practical importance: electricity, cars, mass production factories and the like, derives from the applications of Western science. Without Western technology China would still be where she was 100 years ago, that is in conditions relatively little different from those of 1,000 years ago. The impact of the West has been so great that the Chinese have forgotten even those parts of their own culture which, in certain respects at least, is better than that imported from the 'fair haired, big nosed devils'. One of the very few almost exclusively indigenous discoveries that surpasses its Western equivalent in several respects, is acupuncture. There are many diseases, or physiological dysfunctions, which do not yet amount to a disease, that can be cured by acupuncture and not by Western medicine. Naturally, there are also diseases that can be cured by Western medicine which are intractable to the acupuncturist. The Chinese people themselves often do not sufficiently value the traditional skill of the acupuncturist. They take it too much for granted, much as we in the West take for granted the services of electricity, piped water or rubbish collection, only realising their value when a strike or other action interrupts their functioning.

In this book the viewpoint moves backwards and forwards between traditional acupuncture and scientific medicine. Acupuncture is for the most part based on observed facts which have been woven into a fairly complete system of medicine by a system of theories. The theories themselves are often surprisingly accurate at least insofar as concerns clinical treatment. Not infrequently, however, the theory has been based on philosophical and mystical speculation and can then, often, only be useful as a thread which the mind follows as it weaves together the multitudinous and seemingly isolated threads of factual observation.

Much of what is factually observed in acupuncture could be explained in a way completely different from that of the traditional Chinese account. One example might be a different account by way of the discipline of neurophysiology.

In the chapters that follow I have mentioned nearly all the traditional Chinese theories as I think it is important that they should be known and understood before one starts one's own research. Furthermore, I suspect that much of what seems mystical nonsense to some, in reality portrays many of the laws of nature (even where these are as yet unknown to us) despite the fact that they are expressed in a language that we might call unscientific.

Some doctors or patients may indeed wonder how one can practise a form of medicine where the theories on which that practise is based are possibly suspect. Just as a doctor will prescribe aspirin because he knows what are its effects in the body of a patient, so an acupuncturist will needle a certain acupuncture point because he knows what the consequent reaction of the body will be. It is of secondary importance to the doctor to know just why it is that aspirin has its specific effects, no matter how intellectually interesting such knowledge might be. At the time of writing little is understood of why the known effects of aspirin take place, yet aspirin, with its simple chemical formula, is the most commonly used drug in the world.

The reader will be made aware by various remarks throughout this book, particularly those in chapter XI, that I believe neither in the major part of the traditional Chinese theoretical explanation of acupuncture nor even in its practical application where this is based *solely* on traditional theory. Doctors who follow my courses in acupuncture will find that this divergence in both theory and practice is no hindrance to the successful treatment of a large number of diseases occurring in their patients. Doctors who wish to study acupuncture are welcome to write to me. From time to time I give courses, largely of a practical nature, during which I concentrate on those aspects of the subject that would be difficult to describe in a book.

NEURAL THEORY OF THE ACTION OF ACUPUNCTURE

In acupuncture, the needle is frequently placed at the opposite end, and possibly opposite side, of the body from that of the diseased organ or site of symptoms. Under certain conditions one of these distant and contralateral pricks can have an effect in one or two seconds. This speed of conduction excludes the blood and lymphatic systems (at least in this type of response) and leaves to my way of thinking, the nervous system as the only contender.

There are other, though non-neural, theories:

Kim Bong Han* described a special conducting system of Bong Han ducts and corpuscles, corresponding to the course of acupuncture meridians. Kellner† has shown that some of the above theory is based on artefacts occurring in the preparation of histological slides. Some have thought that the meridians look like the lines of force round a magnet and postulate a magnetic theory. Others somehow manage to bring in quantum mechanics. A Japanese researcher thinks that there is a contraction wave following the course of meridians, along the surface of the skeletal muscles. Some liken the pinprick in the body to the electrical discharge of a condenser. A few say the pinprick releases cortisone or histamine or adrenaline but fail to explain the specific action of the acupuncture points. I once had a theory concerning the lateral line system in fish,‡ which I have since discarded. I am now fairly convinced that the nervous system is the transmission system used in acupuncture. The remainder of this chapter discusses this neural acupuncture theory: part is based on well-known anatomy and physiology, part is conjecture, and part requires experimental proof.

Cutaneo-Visceral Reflex

Acupuncture is based on the fact that stimulating the skin has an effect on the internal organs and on other parts of the body, a rela-

---

*Kim Bong Han. On the kyungrak system. 1964, Foreign languages publishing house, Pyongyang.

†International acupuncture conference in Vienna and German acupuncture conference in Wiesbaden.

‡See chapter XII of the 1st edition of this book.

tively simple reflex whose therapeutic application is largely ignored
in the West. Various experiments demonstrate the existence of this
cutaneo-visceral reflex:

Kuntz, Haselwood; Kuntz; Richins, Brizzee*, in several series of
experiments stimulated the skin on the back of rabbits or rats and
found changes in the duodenum or other parts of the intestinal tract
corresponding to the dermatone stimulated.

By employing a quick freeze-drying technique, it was shown
that when a cold beaker of ice was applied to the back in the lower
thoracic region, the arterioles in the subserosa and submucosa of
the duodenum were constricted and the capillary beds in the villi
were ischaemic. The vascular changes in the subserosa could also
be observed *in vivo* photographically and by plethysmographic
recording.

Reflex responses of the gastric musculature and the pyloric
sphincter in man have been described by Freude and Ruhmann†
using a fluoroscope by means of warm, cold, chemical or mechanical
stimulation of the skin of the epigastrium. They also produced
hyperaemia of the ascending colon after it had been exposed at
operation, by applying heat to the skin of the lower abdominal wall.

Nine patients with angina pectoris or acute myocardial infarction
were investigated by Travell and Rinzler.‡ They found that if the
trigger areas on the front of the chest were infiltrated with procaine

---

*Kuntz, A., and Haselwood, L. A. Circulatory reactions in the gastro-intestinal
tract elicited by local cutaneous stimulation. American Heart Journal, 1940, 20:
743-749.

Kuntz, A. Anatomic and physiologic properties of cutaneo-visceral vasomotor
reflex arcs. Journal of Neurophysiology, 1945, 8: 421-429.

Richins, C. A., and Brizzee, K. Effect of localized cutaneous stimulation on
circulation in duodenal arterioles and capillary beds. Journal of Neurophysiology,
1949, 12: 131-136.

†Freude, E., and Ruhmann, W. Das thermoreflektorische Verhalten von Tonus
and Kinetik am Magen. Zeitschrift für die gesamte experimentelle Medizin, 1926,
52: 338.

Ruhmann, W., Viscerale Schmerzlinderung durch Wärme als Segment-reflex.
Zeitschrift für die gesamte experimentelle Medizin, 1927, 57: 740.

Ruhmann, W. Örtliche Hautreizbehandlung des Magens und Ihre physiologis-
chen Grundlagen. Archiv für Verdaungskrankheiten, 1927, 41: 336.

‡Travell, J., and Rinzler, S. H. Relief of cardiac pain by local block of somatic
trigger areas. Proceedings of the Society for Experimental Biology and Medicine,
1946, 63: 480-482.

or cooled with ethyl chloride, complete and prolonged relief of pain usually ensued.

Of the first series of experiments mentioned above using rats and rabbits, some were performed on animals with an intact nervous system under general anaesthesia, while others were performed on animals where the spinal cord had been transected in the lower cervical region. There was no difference in the two types of experiment, suggesting that the cutaneo-visceral reflex is mediated on a segmental and intersegmental level and not influenced suprasegmentally.

Wernøe* has made similar experiments on fishes and amphibians. In the pithed eel (with its segmental structure), stimulation of 1 sq. cm. of skin with silver nitrate caused, after a delay of two minutes, vasoconstriction of the (from the dermatome point of view) appropriate part of the intestine, followed by a concentric contraction of the intestinal segment, and finally peristalsis, after which bowel movements ceased. If a more proximal or distal part of the skin was stimulated the corresponding sharply defined part of the gut showed the above cycle of events. In the cod electrical stimulation of the skin just distal to the pectoral girdle caused ischaemia of the stomach, whilst stimulation of the skin 5 cm. distal produced ischaemia in a section of the intestine.

Three eels were taken and their brains destroyed. In addition, in the first the entire spinal cord was destroyed, in the second the distal half was destroyed and in the third the spinal cord was left intact. All the skin of the three eels was stimulated with silver nitrate. After a latent period of 2 to 4 minutes the intestine of the first eel showed vasoconstriction, that of the third eel vasodilation, while the second showed vasodilation in the proximal half and vasoconstriction in the distal half. That is: in those sections where the spinal cord is intact there was vasodilation, while in those where it was destroyed, vasoconstriction. In another eel the spinal cord was divided into several segments, of which alternating segments of the spinal cord were destroyed. On stimulating all the skin with silver nitrate, it was again found that the segments with an intact spinal cord had vasodilation whilst the others had vasoconstriction.

From the above Wernøe deduced that the vasodilation was

*Wernøe, T. B. Acta Physiologica Scandinavica.

mediated by a spinal reflex, whilst the vasoconstriction by a post-ganglionic sympathetic reflex.

Viscero-Cutaneous Reflex

The cutaneo-visceral reflex mentioned in the preceding section is of prime importance in acupuncture, for it is by its mediation that an acupuncture needle placed in the correct part of the skin is able to affect the appropriate organ or diseased part of the body.

The viscero-cutaneous reflex to be discussed in this section is of importance (1) in diagnosis and (2) in lowering the threshold of stimulation required in treatment by acupuncture.

It is often noticed clinically that a disease of an internal organ will produce in some part of the skin (not infrequently of the same derma-tome) pain, tenderness, hyperaesthesia, hypoaesthesia etc. This can also be seen experimentally:

Wernøe stimulated the rectum of a decapitated plaice electrically, or with copper sulphate or barium chloride. In each case the skin became pale, due to retraction of the melanophores, extending to 3 or 4 spinal segments of the appropriate dermatomes. Likewise in the eel and cod, the stomach, intestines, gall bladder or spleen were stimulated mechanically or by the intramural injection of 10% adrenalin; in each case the skin becoming lighter over an area of several segments, again in the expected dermatomes.

In the decapitated cod, if the spinal cord is in addition destroyed, the viscero-cutaneous reflex is not abolished. If, on the other hand, the cord is intact but the sympathetic chain is excised the reflex is abolished. This suggests that it is not a spinal reflex, but that it is mediated along unknown paths of the sympathetic chain.

The viscero-cutaneous reflex discussed here and the viscero-motor reflex to be discussed later are presumably the mechanism whereby diseases of internal organs produce pain or tenderness of certain acupuncture points, areas of skin, or muscle spasm. Presumably other, though similar reflexes are involved when diseases other than those of internal organs likewise cause pain, tenderness, muscle spasm etc.

In acupuncture a considerably smaller stimulus is needed if the acupuncture needle is put directly into one of these tender or painful areas or into a meridian that crosses or is related in some other way to the tender area. Presumably facilitation is taking place. This facilitation is also a safeguard, for if the acupuncture needle is put in the wrong place it has little effect, as it is easier to affect a diseased or disease related area than a healthy one.

Viscero-Motor and Viscero-Visceral Reflexes

These reflexes are in many instances similar to the viscero-cutaneous reflexes mentioned above, occurring mostly at the same time, though requiring an intact spinal cord.

Miller, Simpson* and many others stimulated the viscera and obtained muscle contractions in the expected appropriate dermatomes (and distant dermatomes—see later). Distension of the stomach by air, traction on the stomach, mustard oil on the gastric mucosa, squeezing the small intestine, mechanical stimulation of the kidney or spleen all elicited the reflex, which could be abolished by dividing the appropriate dorsal root. The reflex from the stomach was stopped by dividing the splanchnic nerves, while stimulation of the central ends of the divided nerve restored it; similar results were obtained with the superior mesenteric nerves for the small intestine, the hypogastric nerve for the kidney, and the splenic nerve for the spleen.

Brown-Sequard† relates an experiment with a dog which had a tube tied into its ureter. When the internal abdominal wall was pricked within the distribution of the 1st lumbar nerve, the secretion of urine was considerably diminished.

Brown-Sequard† also quotes the case of a colleague, Sir Benjamin Brodie, who had a patient with a stricture of the urethra causing pain and lameness of the foot. All symptoms were relieved after a bougie had been passed up the urethra. The urethra and foot symptoms were probably in the same or neighbouring dermatome. I have on several occasions observed in a patient with multiple sclerosis that when the heel is pricked with a needle the patient micturates.

Cannon and Murphy‡ compared two series of cats who had been given ether anaesthesia. The one group had their testicles crushed whilst the others were not molested. Afterwards both groups were

*Miller, F. R., and Simpson, H. M. Transactions of the Royal Society of Canada, sec V, 1924, XVIII, 147.

†Brown-Sequard, C. E., Course of lectures on the physiology and pathology of the central nervous system. 1860, Lippincott, Philadelphia p. 170 and 167.

‡Cannon, W. B., and Murphy, F. T. Physiologic Observations in experimentally produced ileus. Journal of the American Medical Association, 1907, 49: 840.

given food mixed with barium and its passage in the ileum followed roentgenologically. In the emasculated group there was no movement of the ileum for 4 hours, after which it started sluggishly; whilst in the control group there was movement from the very beginning. The testicles are usually given as T10 to L3 (varying according to the author) and the small intestine as T6 to T11, so that this is possibly a segmental reflex.

Dermatomes and Acupuncture Points

Some of the acupuncture points, particularly those on the back, having an effect on a specific organ, are in the appropriate dermatomes as a glance at Fig. 1* and 2 will show. Account should be

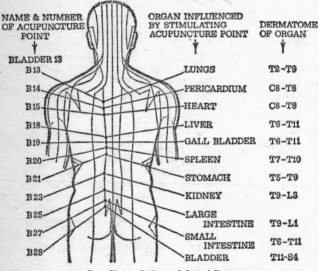

NAME & NUMBER OF ACUPUNCTURE POINT	ORGAN INFLUENCED BY STIMULATING ACUPUNCTURE POINT	DERMATOME OF ORGAN
BLADDER 13		
B13	LUNGS	T2-T9
B14	PERICARDIUM	C8-T8
B15	HEART	C8-T8
B18	LIVER	T6-T11
B19	GALL BLADDER	T6-T11
B20	SPLEEN	T7-T10
B21	STOMACH	T5-T9
B23	KIDNEY	T9-L3
B25	LARGE INTESTINE	T9-L1
B27	SMALL INTESTINE	T6-T11
B28	BLADDER	T11-S4

FIG. 1 Hyperaesthetic zones in internal disease

*Dermatomes derived from Hansen, K. and Schliak, H. Segmentale innervation, ihre bedeutung für Klinik und Praxis, 1962, Georg Thieme, Stuttgart; and other sources.

taken though that the Chinese do not always mean the same as we
do when they say heart, liver etc. as described in greater detail in
'The Meridians of Acupuncture'.

The majority of reflexes mentioned in the previous section

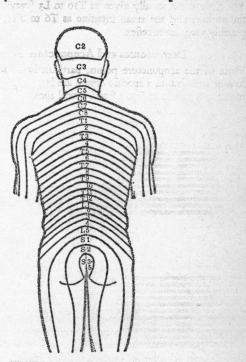

FIG. 2 Dermatomes, after Keegan and Garrett

(cutaneo-visceral, viscero-cutaneous, viscero-motor) are segmental in
nature, and hence fit in with the dermatome pattern of acupuncture
points described in this section. Some of the reflexes can also be
intersegmental, not following the dermatomes. These are described
in the next section.

There is often considerable variation in the dermatome pattern according to the method of investigation: hyposensitivity from loss of function of a single nerve root (Keegan and Garrett);* electrical skin resistance in sympathectomised patients; electrical skin resis-

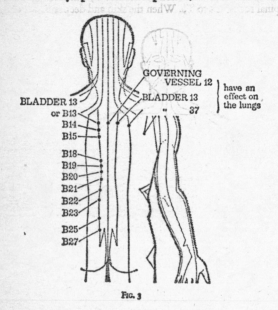

GOVERNING
VESSEL 12

BLADDER 13
or B13

B14
B15

B18
B19
B20
B21
B22
B23

B25
B27

BLADDER 13

37

} have an
effect on
the lungs

FIG. 3

tance on stimulation of anterior spinal roots; pain distribution after hypertonic saline injection of interspinous ligaments (Kellgren).

There is also variation between individuals: Sixteen patients were examined to determine the electrical skin resistance of the arm at operation,† by stimulating the anterior spinal roots. The upper limit

*Keegan, J. J., and Garrett, F. D. The segmental distribution of the cutaneous nerves in the limbs of man. Anatomical Record, 1948, 102: 409–439. Also Fig. 2 and 12.

†Ray, B. S., Hinsey, J. C., Geohegan, W. A. Observations on the distribution of the sympathetic nerves to the pupil and upper extremity as determined by stimulation of the anterior roots in man. Annals of Surgery, 1943, 118, No. 4: 647–655.

varied from T1 to T3, whilst the lower limit varied from T7 to T10. The usual range is T2 to T9.

Normally the dermatome of the arm is given as C5 to T1, whilst the sympathetic dermatome obtained by stimulating the anterior spinal root is T2 to T9. When the skin and deeper tissues are pierced

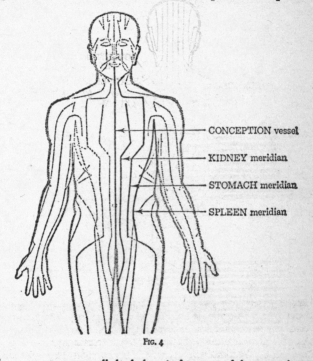

CONCEPTION vessel

KIDNEY meridian

STOMACH meridian

SPLEEN meridian

FIG. 4

by an acupuncture needle both the spinal nerves and the sympathetic nerves of the blood vessels are affected, that of the sympathetic nerves having no effect unless one believes in antidromic stimulation.

The series of acupuncture points on the back lateral to the associated points shown in Fig. 3 have an effect on the same organ as its sister point at the same level, e.g. both B13 and B37 influence the lungs, (likewise the points on the governing vessel).

The abdomen and front of the chest are traversed by the spleen, stomach, kidney and conception meridians (Fig. 4), which despite their names have, I find, an effect mainly on the region of the body traversed. All four meridians where they cross the chest may be used to treat the lungs and heart, yet their abdominal course influences the abdominal viscera. Thus as with the acupuncture points on the back a rough dermatomal pattern is preserved.

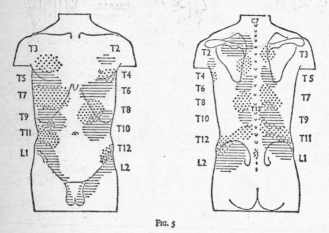

Fig. 5

It is interesting to note that the majority of acupuncture points on the abdomen, thorax and back are near the mid-ventral and mid-dorsal lines. This corresponds to the segmental reference of deep pain obtained by Kellgren* when injecting hypertonic saline into the interspinous ligaments (Fig. 5).

The meridians of the lung, pericardium and heart on the anterior surface of the arm correspond at least approximately to the appropriate dermatome (Fig. 6). The meridians on the posterior surface of

*Kellgren, J. H. On the distribution of pain arising from deep somatic structures with charts of segmental pain. Clinical Science, 1939-42, 4: 35-46. Also Fig. 5.

the arm: the large intestine, small intestine and triple warmer, do not correspond to the appropriate dermatome (Fig. 7). It should be noted though that, despite their name, stimulation of a large intestine acupuncture point influences the lung more than the large intestine and stimulation of a small intestine point influences the heart more than

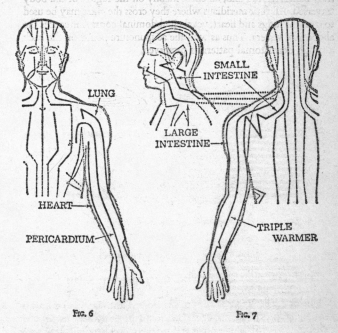

LUNG

SMALL
INTESTINE

LARGE
INTESTINE

HEART

PERICARDIUM

TRIPLE
WARMER

FIG. 6   FIG. 7

the small intestine; whilst stimulation of the triple warmer produces effects hard to catalogue under a single organ. The association in Chinese theory of the lung and large intestine on the one hand, and the heart and small intestine on the other is explained later in this book. In the context of the present discussion the association would mean that the stimulation of either lung or large intestine acupuncture points has an effect primarily on the lung, and only secondarily influences the large intestine; whilst the stimulation of either heart or small intestine acupuncture points has an effect primarily on the heart, and only secondarily influences the small intestine. Possibly this is similar to distension of the stomach secondarily causing displacement of the heart.

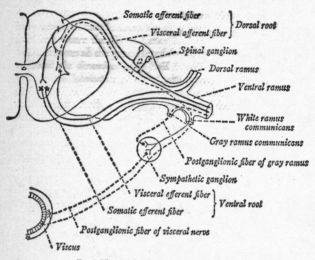

FIG. 8 The peripheral nerves and spinal cord

In this section we have discussed the acupuncture points that more or less fit in with a dermatological pattern (Fig. 8),* namely—the whole of the back, the abdomen, the front of the chest and the arms. The acupuncture points of the legs and head do not fit in with what is known of dermatomes and are therefore described in relation to other neurological concepts below:

Intersegmental Reflexes and Acupuncture Points

The acupuncture points on the legs are those of the liver, gall bladder, kidney, bladder, spleen and stomach. In all cases (except the bladder)

*From Ranson, S. W. and Clark, S. L. The anatomy of the nervous system, 1966, Saunders, Philadelphia.

the dermatomes of these organs are on the trunk and not on the legs. It is however an undoubted fact, observed every day by any doctor who practises acupuncture (for the leg acupuncture points are commonly used), that stimulation of a leg acupuncture point does

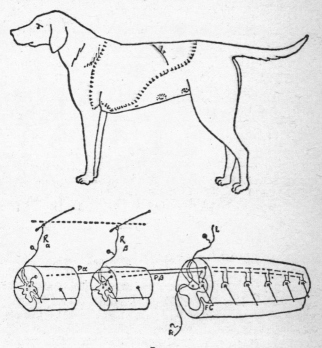

FIG. 9

have an effect on the appropriate organ, even though it may be ten dermatomes away. A possible explanation is via intersegmental reflexes, called by Sherrington long reflexes, whilst those effects of acupuncture that fit in with the dermatomes are segmental reflexes — Sherrington's short reflexes.

Sherrington* described the *scratch reflex* in the spinal dog (Fig. 9) in which stimulation anywhere in a saddle-shaped area extending from the pectoral to the pelvic girdle caused rapid scratching movements in the ipselateral hind leg and rigidity in the contralateral limb. If the stimulus is moved but slightly to the opposite side of the back the hind legs reverse their roles. Ipselateral hemisection of the spinal cord abolishes the reflex, contralateral hemisection leaves it unaffected.

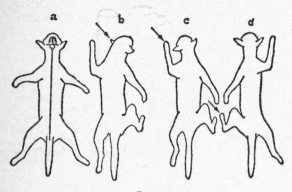

FIG. 10

Sherrington also experimented with decerebrate cats in which the nervous axis is divided at the level of the mid-brain. In the resultant decerebrate rigidity, the cats exhibit *reflex figures* (Fig. 10).

(*a*) In normal decerebrate rigidity all limbs are extended.

(*b*) If the left pinna is stimulated there is flexion of the left fore and right hind limbs, with increased extension of the others.

(*c*) If the left fore limb is stimulated there is flexion of the left fore and right hind limbs, with increased extension of the others.

(*d*) If the left hind limb is stimulated there is flexion of the left hind limb and right fore limb, with increased extension of the others.

The reflex figures require both sides of the spinal cord for their

*Sherrington, C. S. The integrative action of the nervous system, 1906, Scribner, New York. Also Fig. 9 and 10.

conduction, not only the one as in the scratch reflex. Both the scratch reflex and reflex figures are intersegmental (jumping several dermatomes) cutaneo-motor reflexes.

Downman* investigated long viscero-motor and long cutaneo-motor reflexes in the cat with a spinal transection at T1. The splanchnic nerve serving the viscera, intercostal nerves T3–T13, lumbar nerves L1–L3 and the tibial nerve at the knee were all exteriorised.

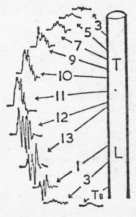

FIG. 11

Maximal single-shock stimulation of the central end of the splanchnic nerve evoked reflex volleys in all body wall nerves and the tibial nerve (Fig. 11). Even at threshold stimulation several intercostal nerves were involved. If an intercostal nerve was stimulated the response was in some cases as large as with splanchnic stimulation.

Downman showed that splanchnic excitation can spread up the

*Downman, C. B. B. Skeletal muscle reflexes of splanchnic and intercostal nerve origin in acute spinal and decerebrate cats. Journal of Neurophysiology, 1955, 18: 217–235. Also Fig. 11.

Downman, C. B. B., and McSwiney, B. A. Reflexes elicited by visceral stimulation in the acute spinal animal. Journal of Physiology, 1946, 105: 80–94.

cord by (1) a fast extraspinal route in the sympathetic chain of the same side and (2) a slower intraspinal route of limited ascent. Intercostal excitation can ascend only by a slow intraspinal route. This was demonstrated by the following experiments: Reflex discharges into the lower intercostal nerves on both sides were elicited by stimulating the left splanchnic nerve. Cutting the left sympathetic chain limited the upward spread of the excitation to the next 3 to 5 segments of the cord. The discharges in the nerves were now of decreasing size and of longer latency in these segments. Spread of activity on stimulating a lower left intercostal nerve was unaffected. Where the chain had been left intact and the cord transected, splanchnic excitation spread freely into segments above the transection, but spread of intercostal excitation stopped at this level. In those instances where there is a contralateral response, experiments involving unilateral section of the dorsal nerve roots were performed. It was concluded that the splanchnic afferent volleys enter the cord by the dorsal root, traverse the spinal cord and leave by the contralateral intercostal nerves. Similar research has been done by Miller, Ward*, and Duda.†

There are also long intersegmental viscero-visceral reflexes. The gastric-colic reflex is invoked when food entering the stomach causes mass contractions of the colon. Likewise in travel sickness where the afferent fibres are the trigeminal, glossopharyngeal or vagus and the efferents are the phrenics and intercostal nerves.

I have thus been able to show in this section that the leg muscles contract if, under the correct conditions, one stimulates: the abdominal viscera, the splanchnics, the intercostal nerves, the outer ear, the front feet, the skin of the back in the upper thoracic region and other areas many segments away from the dermatomes of the human leg (or hind leg in animals). The reverse of the above, namely stimulating the skin of the leg having an effect on the viscera, was demonstrated by Brown-Sequard in the same course of lectures mentioned earlier. He poured boiling water over the hind leg of a

*Miller, F. R., and Ward, R. A. Viscero-motor reflexes. American Journal of Physiology, 1925, 73: 329–340.

†Duda, P. Facilitatory and inhibitory effects of splanchnic afferentation on somatic reflexes. Physiologia Bohemoslovenica, 1964, 13: 137–141.

Duda, P. Localization of the splanchnic effect on somatic reflexes in the spinal cord. Physiologia Bohemoslovenica, 1964, 13: 142–147.

dog whose spine was divided at L3 and another dog whose spine was divided at T3. At autopsy two days later the former dog showed congestion of the bladder and rectum (segmental), whilst in the latter all abdominal organs were congested (intersegmental).

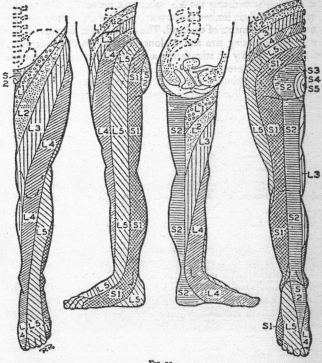

FIG. 12

The distribution of the acupuncture points on the legs is such that each organ corresponds to several dermatomes and each dermatome corresponds to several organs.

This problem can be partly resolved when it is realised that the acupuncture points on the medial and anterior side of the thigh, (liver, spleen, kidney and stomach meridians) do not have very much effect on the liver, spleen, kidney or stomach, as the name suggests, but affect mainly the reproductive organs. There are similar discrepancies with some other points.

If the dermatological charts of Keegan and Garrett, obtained by charting the hyposensitivity from loss of a single nerve root, are taken (Fig. 12) the remaining acupuncture points fit more easily into a dermatome pattern. The kidney and bladder, which from an acupuncture point of view function together would be S1 and S2; the gall bladder and stomach L5; the liver and spleen, which are hard to distinguish, L3 and L4. Perhaps the long intersegmental reflexes for the legs do not follow a dermatological pattern.

The problem is not too simple, as investigations by Travell and Bigelow* showed. In patients with pain encompassing several dermatomes it was found that a pinprick to the trigger area might relieve the pain in that dermatome or in several dermatomes or in one dermatome, then miss out a dermatome to relieve pain again in a further dermatome.

Acupuncture Points on the Head — Near and Distant Effects

Most of the acupuncture points on the head have a local effect, which could presumably be explained by local reflex arcs similar to segmental reflexes.

The apportioning of the acupuncture points on the head to the various internal organs is hard to follow both theoretically and in actual clinical acupuncture practice, though distant effects undoubtedly occur.

Koblank† investigated a reflex between the nose and the heart. He found a sharply defined area in the region of the superior concha of the nose, which if stimulated with a probe caused various cardiac arrhythmias in man, dogs and rabbits. When the vagus was cut on one side, the reflex remained intact; when cut on both sides the reflex was abolished for a few days and then returned, but weaker than formerly. When the maxillary nerve was divided on one side

*Travell, J., and Bigelow, N. H. Referred somatic pain does not follow a simple segmental pattern. Federation Proceedings, 1946, 5: 106.
†Alfred Koblank. Die Nase als Reflexorgan. 1958. Haug, Ulm. Also Fig. 13.

the reflex was permanently abolished when the same side of the nose was stimulated, but the reflex persisted normally when the healthy side was stimulated. From this it was deduced that the trigeminal nerve relayed the stimulation of the nasal mucous membrane to the region of the nucleus of the vagus, which then passed it on via the vagus to the heart. It was considered though that there was more than one final pathway as dividing the vagus only partially abolished the reflex.

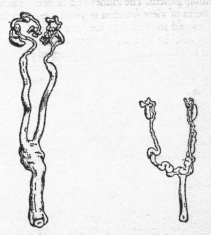

FIG. 13 Left: control. Right: after excision of inferior turbinate

Koblank also investigated the relation between the lower turbinate of the nose and the reproductive organs of rabbits and dogs. He found that if the lower turbinate was excised in young animals that the uterus, fallopian tubes, ovary or testicle failed to develop, even though the adult animal attained the same weight as an unoperated control. The failure of development showed itself both as a considerable reduction in size (Fig. 13) and histologically.

Koblank, Röder and Bickel experimented with dogs who had a Pavlov type exteriorised blind loop, whereby changes in gastric secretion and motility could be observed directly. They found that when the 'stomach area' on the anterior third of the middle turbinate was stimulated that the gastric secretion and movement were increased.

It should be noted that in the above experiments stimulation of the upper turbinate affected the heart, the middle turbinate the stomach, and the lower turbinate the reproductive organs.

Specific Response versus Generalised Response

In the practice of acupuncture it is sometimes found that one (or a small group) of acupuncture points are effective in treating a certain patient. On other occasions, any one of several meridians (encompassing a large number of acupuncture points) can be effective. In the former case a specific stimulus is mandatory, in the latter nearly any general stimulus is all that is needed.

The specific response presumably takes place, along the lines of the nervous pathways described in the previous sections.

The generalised hypersensitivity on the other hand seems similar to the pain one can sometimes have with severe toothache when the whole of the same side of the face, arm and upper chest are hypersensitive. In the same way the viscera may sometimes become hypersensitive affecting the nerves in a large area, and hence only require in treatment an acupuncture needle put anywhere in a large area, in any of a large number of acupuncture points, or in any of several meridians.

In other cases a stimulus anywhere in a large area does not depend on hypersensitivity, but on the large number of neurones that have a final common path. Ashkenaz[*] stimulated the gall bladder of cats by inflating a balloon. This caused contraction of the panniculus carnosus muscle (the cat's equivalent of the platysma, but extending over most of the body). This viscero-pannicular reflex was only abolished when all the dorsal roots T2 to T9 were severed, a single root being sufficient to preserve the reflex, thus demonstrating the convergence that can take place.

Diseased organs seem to have a lowered threshold of response, for only a small stimulus is needed to correct a dysfunction of a

[*]Ashkenaz, D. M. An experimental analysis of centripetal visceral pathways based upon the viscero-pannicular reflex. American Journal of Physiology, 1937. 120: 587–595.

diseased organ. On the other hand a very considerable stimulus is needed to alter the function of a healthy organ. For this reason the small prick of an acupuncture needle can cure some of the severest diseases, and yet is normally harmless if the wrong treatment is effected, as the threshold of response of the healthy organ is beyond the stimulus of a mere needle prick.

It should be noted that the Chinese describe the acupuncture points as being quite small—a matter of millimetres. In my experience this is only true to a limited extent, for not infrequently a stimulus anywhere in an area as large as a dermatome (or several dermatomes if there has been spread of hypersensitivity) is sufficient. If this largish area is carefully examined by hand a few small areas of maximal tenderness, with possibly small fibrositic-like nodules, will be found (similar to the small areas of maximal tenderness found when a large area such as the neck and shoulders are 'rheumatic'). If these small areas of maximal tenderness, or the 'fibrositic' nodules are stimulated by an acupuncture needle the response is normally greater than when the surrounding less tender area is needled. If the dysfunction of a diseased organ is mild, a reflex tenderness may not be produced over the whole of a dermatome, being demonstrable only in a few small tender areas—the same areas as mentioned a few lines above. These small tender areas of 'fibrositic' nodules are relatively constant in position, whether the remaining surrounding part of the dermatome is tender or not. This constancy in position applies from one individual to another, and is likewise the same for any variety of diseases producing a reflex tenderness in that area. It is these small tender areas of constant position, which are termed the acupuncture points; although, as mentioned above, a stimulus anywhere in the appropriate dermatome (or sometimes even larger area) may work, albeit frequently not so well.

Some years ago David Sinclair, professor of anatomy at Aberdeen University, wrote an as yet unpublished paper, which he has kindly let me read, concerning the reflexes between the skin and viscera defined as viscero-somatic, somato-somatic, viscero-visceral, somato-visceral. In this article Sinclair quotes a hundred papers (several of which are mentioned in this chapter) concerning these reflexes which are the presumed mechanism of acupuncture—though probably most of the authors know nothing or little of acupuncture. At the time of writing and in other papers* Sinclair advanced a branched axon theory partially to explain the observed phenomena, but since then he thinks the more conventional nervous pathways are the mediator.

I think the neurophysiological theory to explain the mechanism of acupuncture, which I have developed over the past years and described in this chapter, will soon be recognised as the basis for the scientific investigation and further development of acupuncture. No doubt there will be vast extensions, modifications and contradictions. But I will be glad if my investigations have sown a seed that others – neurophysiologists and clinicians – may tend further.

*Sinclair, D. C. The remote reference of pain aroused in the skin. Brain, 1949, 72: 364.

Sinclair, D. C., Weddell, G., and Feindel, W. Referred pain and associated phenomena. Brain, 1948, 71: 184.

II

THE ACUPUNCTURE POINTS

In all diseases, whether physical or mental, there are tender areas at certain points on the surface of the body, which disappear when the illness is cured.

These are the so-called acupuncture points.*

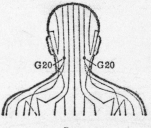

FIG. 14

In some cases they will be spontaneously painful. When, for example, a patient suffers from a frontal headache, he will feel pain just lateral to the upper end of the trapezius, a point known as gall bladder 20 (Fig. 14). In other cases, however, these points are only tender under pressure, so that there would be no pain at all at, say,

*The anatomical position of the acupuncture points may be seen in my 'Atlas of Acupuncture'.

gall bladder 20 until pressure was applied. To this category belong
the many points just above the ankle. Women in particular are
unaware how tender this area is unless it is pressed. Thirdly there is
the type of acupuncture point where there is no tenderness at all,
even under pressure, so that they can only be found by the hand of
the experienced physician.

The doctor looking for an acupuncture point will, in the simplest
of instances, discover a little nodule, like the fibrositic rheumatic
nodules often present at the back of the neck, in the shoulders or in
the lumbar area (Fig. 15 upper). But in many cases instead of the
nodule the examining finger may find a strip of tense muscle within

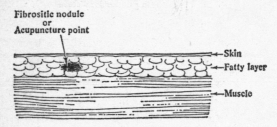

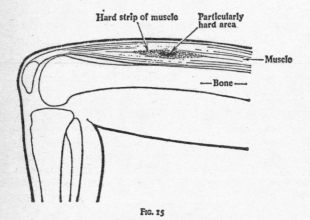

FIG. 15

a group of muscles with a particularly hard and indurated area (Fig. 15 lower). Sometimes there is an area which is slightly swollen or discoloured. In the most difficult cases the point cannot be found without a knowledge of its exact anatomical position.

Some people use an electrical instrument to measure the electrical

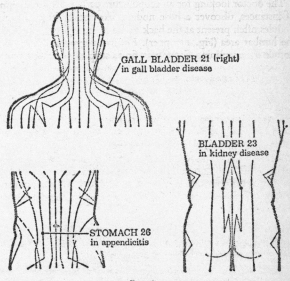

FIG. 16

skin resistance or impedance. The theory is that, since the impedance is reduced at the acupuncture points, they can in this way be easily and accurately discovered. I have myself tried several types of experimental apparatus, but have found that the electrical resistance both in living bodies and cadavers varies in so many places, not only at the acupuncture points but thousands of others, that to me this apparatus is not of much use. If the special electrical properties of the nervous system are taken into account, it should be possible to measure some types of electrical variation.

As mentioned above, the pain at those acupuncture points which are either spontaneously tender or tender under pressure vanishes with the cure of the disease. It makes no difference how the cure has been achieved, whether by acupuncture, ordinary drugs, osteopathic manipulation, homeopathy, hypnosis or the mere passage of time: when the illness ends, so also does the pain.

This at once establishes a causal relationship between disease, physical as well as mental, and the tender variety of acupuncture point.

In one very simple form of acupuncture diagnosis, the patient is examined from head to toe in order to find all the tender points, and hence to deduce the internal disease corresponding to them. Some of these points are known and used for diagnosis in orthodox medicine, though of course the average doctor is unaware that these are acupuncture points. Thus the right shoulder, particularly at point gall bladder 21, may be spontaneously tender in gall bladder disease, bladder 23 in renal disease and McBurney's point near point stomach 26 in appendicitis (Fig. 16).

The acupuncture points can serve a dual purpose, for not only do they help in the diagnosis of disease but may also conversely be used for its treatment. In this the skin is pierced at the acupuncture point by a fine needle, which is withdrawn usually after the lapse of a few minutes.

When the vision is blurred and the eye does not see, the side of the head is painful, likewise the outer corner of the eye. This is cured by needling the point 'Jaw Detested' (gall bladder 4).

(Jia Yi Jing, Vol. XII, Ch. 4)

In one form of acupuncture the points spontaneously tender or those tender under pressure are needled. In other more refined forms, the acupuncturist needles those points where no pain is felt at all, points which are often remote from the seat of the disease and sometimes even on the opposite side of the body.

Occasionally some wholly accidental stimulus to an acupuncture point may cure a disease. I remember when I was at school seeing a couple of boys fighting together on top of a bed. One of them fell down, hitting his forehead at the root of the nose on the iron bedstead, and was immediately cured of the sinus trouble he had suffered from for two or three years. A parallel case is that of a woman with a dull, though mild, headache accompanied by general malaise, which persisted day after day almost unrelieved for some ten years. During these years she had had several blood tests, which involved pricking the skin to reach a vein at the elbow. She soon noticed that, every time her skin was pricked at this point, the headache and malaise instantly disappeared and for a couple of hours she was free of

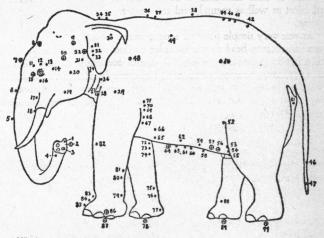

● Nila (nerve centres)

⊕ Nila that kill if prodded hard,
i.e.:—
7, 8, 9, 10, 15, 22, 28, 56, 57,
64, 86, 87, 89, 90, 78, 51

FIG. 17

pain. This happened so regularly that she actually began to look forward to her blood tests; but, when she mentioned this to her doctor, tentatively suggesting a connection between the tests and the relief of the headache, he dismissed the idea as nonsensical. When she came to me, I inserted an acupuncture needle into the same place at the elbow. I neither pierced the vein nor drew off any blood; yet the headache at once disappeared, a proof that the cause and effect she had noticed had nothing to do with the loss of blood. She was in time completely cured of her trouble by the additional needling of several other acupuncture points, effective in her particular type of headache.

There are many such apparently accidental correspondences, some of them not generally known. The knock-out points of Judo, for example, are also acupuncture points, which if too strongly stimulated, will cause the subject to collapse in a faint. The Indian points of the Chakras and Nadir similarly correspond to acupuncture points. Deraniyagala, director of the national museums of Ceylon, lists the places which the *mahout*, or Indian elephant boy, prods with a sharp stick to elicit various responses from his elephant* (Fig. 17). Having no personal experience here, I do not know if these 'Nila' are really acupuncture points, but at least it seems feasible; and perhaps the reason why the African, unlike the Indian, elephant cannot be adequately trained is because the Nila of the African elephant are unknown.

The functions of the Nila are as follows:-

1. Twists trunk
2. Straightens trunk
3. Frightens
4. Frightens and makes trumpet
5. Frightens, makes trumpet and stops animal
6. Brings under control
7. Kills
8. Kills
9. Kills
10. Kills
11. Brings under control
12. Brings under control
13. Rouses
14. Brings under control
15. Kills
16. Kneels
17. Goes backwards
18. Controls animal while being tied to a tree
19. Gives his shoulder
20. Lowers head and neck and stops
21. Brings under control
22. Kills
23. Bends head
24. Stops animal
25. Rouses, infuriates
26. Stops animal
27. Offers seat
28. Kills
29. Stops
30. Brings under control
31. Travels
32. Travels
33. Travels
34. Lowers head
35. Benumbs
36. Stops animal as well as makes animal walk

*Some Extinct Elephants, Their Relatives and Two Living Species (Ceylon National Museum Publication).

37. Stops animal as well as makes animal walk
38. Lowers seat
39. Frightens
40. Frightens
41. Frightens
42. ?
43. Walks
44. Walks
45. Walks
46. Stops animal
47. Travels
48. Stops animal and makes it walk
49. Offers seat
50. Stops without fidgeting and puts trunk to ground
51. ?
52. Gets up and runs
53. Turns round
54. Turns round
55. Turns round
56. Kills
57. Kills
58. Drops to ground
59. Turns round
60. Rouses, infuriates

61. Rouses, infuriates
62. Turns round
63. Rouses, infuriates
64. Kills
65. Stops animal
66. Stops animal
67. Stops animal
68. Stops animal
69. Kneels
70. ?
71. Kneels
72. } Travels when 2 nila are
73. } touched; stops when 1 nila
74. } is touched
75. Raises fore foot for mahout to mount
76. Gives fore foot
77. Gives fore foot
78. ?
79. Brings hind foot forward
80. } Offers hind foot and
81. } twists
82. Draws hind foot backward
83. Raises fore foot
84. Raises fore foot
85. Raises fore foot
86. Kills etc.

Several indigenous medical systems in different parts of the world probably correspond to a simple form of acupuncture. Thus some Arabs will cauterise part of the ear with a red-hot poker in treating sciatica, while among the Bantus of South Africa certain healers will scratch small circumscribed areas of the skin and then rub various herbs into them.

A doctor often has to diagnose mysterious abdominal pains or other symptoms, for which he can find no definite cause. He may therefore suggest to his patient an exploratory operation, in case there is any serious disease present. The surgeon will probably find a spastic colon or mild inflammation of the abdominal lymph nodes or some other not irreversible condition. So he does nothing beyond sewing up the patient and sending him home after a week in hospital.

At the follow up a month later, often the patient will tell his doctor that he has been completely cured of his illness, and thanks the surgeon for his skilled and timely operation. As for the surgeon, he will probably think in some bewilderment that he must have cured his patient by hypnosis.

One surgeon who taught me as a medical student, believed that a little air let into the abdomen cured all manner of ills. But is it not possible that the patient was cured because the surgeon's knife stimulated several acupuncture points? If so, would it not be much simpler to try acupuncture in such cases and leave to the surgeon only those where surgery is really necessary?

Case History. A patient limped into my consulting room with severe pain in the lower back and leg due to a slipped disc. Physiotherapy, a corset and traction had been tried to no avail. The next attempt on the agenda was a major operation to fuse several vertebrae.

I tried acupuncture and the patient was soon cured, but he still has to be careful lifting heavy weights. In some cases where the disc is severely prolapsed, an operation may be the only answer, but everything, including acupuncture, should be tried first.

Case History. A lady in her late thirties had severe menorrhagia (called in Chinese 'bursting and leaking disease'). She was anaemic, weak and often had to stay at home. No treatment had helped and her gynaecologist suggested a hysterectomy. Acupuncture cured her—and she still has her womb today.

If the menorrhagia is caused by large fibroids, a growth or certain other conditions, surgery or radiation is often the best treatment. In the majority of instances menorrhagia is due to a mild dysfunction of the uterus or endocrine system, which can often be cured or helped by acupuncture.

III

THE MERIDIANS

The means whereby man is created, the means whereby disease occurs, the means whereby man is cured, the means whereby disease arises: the twelve meridians are the basis of all theory and treatment.

(Ling Shu, jingbie pian)

In Chinese literature there are descriptions of about a thousand acupuncture points, though there may well be even more than this. Books on the subject are full of accounts of illnesses which can be cured or alleviated by stimulating with a needle one or other of these points. The point called bladder 7 (Fig. 18) for instance, near the top of the head with the picturesque Chinese name of Penetrating Heaven, has an effect on:-

Headache, heaviness of head, glands swollen in the neck, nose blocked, epistaxis, rhinorrhoea, loss of smell due to catarrh, swelling of the face,

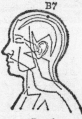

FIG. 18

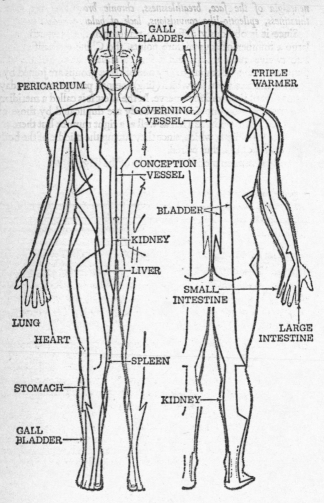

FIG. 19

neuralgia of the face, breathlessness, chronic bronchitis, dry mouth, thirstiness, epileptic-like convulsions, lack of balance, weak eyesight.

Since it is obviously difficult to remember the properties of so large a number of acupuncture points, the Chinese classified them into twelve main groups and a few subsidiary ones. All the acupuncture points belonging to any one of these groups are joined by a line, the Chinese word for which (Jing) means a passage, or nowadays forms part of the word for a nerve. In the West it is called a meridian. The meridians on one side of the body are duplicated by those on the other, just as we have a left as well as a right thumb; but there are two extra meridians, which, since they run up the middle of the body cannot of course be thus paired.

The twelve main meridians are those of the (Fig. 19):—

lung	bladder
large intestine	kidney
stomach	pericardium
spleen	triple warmer
heart	gall bladder
small intestine	liver

(There is also the governing vessel and conception vessel – see later.)

The number of acupuncture points along each of these meridians varies, the heart meridian, for example, having nine points on each side (Fig. 20) while the bladder meridian has sixty seven. All the acupuncture points on a meridian affect the organ after which they are named.

Case History. A patient suffered from recurrent palpitations and a feeling of pressure across the chest, sometimes during periods of slight mental or physical stress, sometimes for no apparent reason. She easily became breathless walking upstairs, had less than her normal energy and was therefore compelled to rest during the daytime, which meant that she found it harder to get through the day's work.

These symptoms were clearly due to a heart condition; so a needle was inserted at the acupuncture point heart 7 (Fig. 20), called by the Chinese the 'gateway of the spirit', since the spirit was thought to live in the heart. Within a few minutes the symptoms were alleviated and, after half a dozen repetitions of the treatment at fortnightly intervals, the patient was cured.

*Taken from: The Treatment of Disease by Acupuncture.

Clearly this system of classifying a thousand acupuncture points into twelve main (and two extra) meridians is in practice very useful, for if, as in the above instance, a patient has a disease of the heart, one immediately knows which group (meridian) of acupuncture

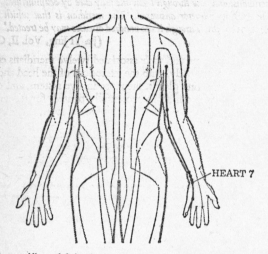

HEART 7

FIG. 20 Heart meridian on left showing its acupuncture points. Heart 1 (under armpit) along the arm to heart 9 (at end of little finger).

points to use. In this case acupuncture point 7 was used, though any of the other eight points on the heart meridian would have helped, but to a lesser extent. The acupuncture point which has the greatest curative effect in a particular disease or on a particular patient is discussed later in the course of this book.

'The kidney meridian starts at the sole of the foot... When it is diseased, the face turns black as charcoal, there is loss of appetite, coughing of blood, harsh panting, a wish to get up when sitting down, the eye cannot see clearly, the heart feels as if suspended.'

(Jia Yi Jing, Vol. II, Ch. 1a)

THE TWELVE SPHERES OF INFLUENCE IN THE BODY

The Thunder God said, 'I would like to know about the course and diseases of meridians, and how through them one may cure by acupuncture.'

The Yellow Emperor answered, 'The meridian is that which decides over life and death. Through it the hundred diseases may be treated.'

(Jia Yi Jing, Vol. II, Ch. 1a)

The twelve organs and their associated twelve meridians encompass all parts of the body, with the exceptions of the head and sense organs, the endocrine glands, the reproductive system, and others. Nevertheless, these can all be successfully treated, for (though not directly counted among them) they all belong to one or more than one of the twelve main meridian groups.

The Main Meridian

Anything that happens along or near the course of a main meridian will influence that meridian and the organ which bears its name.

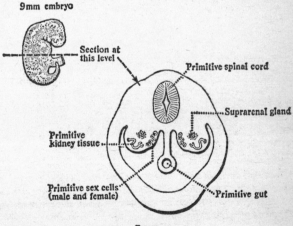

FIG. 21

Case History. A patient who had been troubled with palpitations and breathlessness for two months, came to consult me, thinking she had a disease of the heart. In the course of our conversation, she mentioned that two days before the onset of the symptoms she had sprained her wrist, and I found that the place where the tenderness was most acute crossed the heart meridian at acupuncture point 7 (Fig. 20). The constant irritation at this point over several days had in turn affected the heart. In this instance the patient was cured of her heart symptoms by treating the sprained wrist rather than by a direct treatment of the heart. The similarity between this case history and that a few pages back, which was concerned with a genuine disease of the heart, should be noted.

Embryological Relationships

When acupuncture points on the kidney meridian are stimulated, they affect not only the kidney but also embryologically related organs, such as the ovary, testicle, uterus, fallopian tube and, to some extent, the adrenal. This is because all these organs are formed in the same region of the embryo—the region of the kidney. This intimate relationship in the embryo is maintained in the adult, at least in so far as kidney acupuncture points are concerned (Fig. 21).

This, only one among hundreds of embryological relationships, is an example of how the interdependence of different parts of the body can be utilised in acupuncture.

Anatomical and Functional Relationships

In the adult the nose and throat are part of the respiratory system; in the embryo, however, the throat is part of the alimentary tract, from which the nose is split off. As far as acupuncture is concerned,

LUNG 7

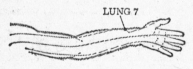

Fig. 22

diseases of the nose and throat can usually be treated through lung acupuncture points, the lung being the main respiratory organ. Hence, in this instance, the adult function predominates rather than the embryological relationship described in the previous section. Nasal catarrh, or hay fever, for example, may be treated in this way by using acupuncture point lung 7 (Fig. 22), though as a rule a few accessory points are also needed.

Physiological Relationships

The stimulation of one of the fourteen paired acupuncture points on the liver meridian will improve an obvious hepatic disease, like jaundice; but several other ailments less evidently from this source can also be cured or alleviated, such as:—

Migraine, a condition which makes the patient feel nauseated and bilious—and indeed was once known as 'the megrims' or bouts of biliousness.

Cyclic vomiting in children; and what is commonly called 'feeling liverish'.

Certain allergic conditions, such as nettle-rash, asthma and hay fever. Some of the antibodies are manufactured in the liver.

Gout, which is a metabolic disease of the liver.

A tendency to bruise easily, presumably because a weak liver will not produce enough prothrombin, or other clotting agents.

Weak eyesight, pain in, behind or round the eyes and black spots or zig-zags floating in front of them. The traditional Chinese belief in the relationship between eyes and liver may explain this condition.

An inability to wake fresh and alert in the morning, however early one has gone to bed, is often due to the liver.

Some weakness or disorder of the liver is commonly (though not invariably) at the root of all these troubles.

Case History. A photographer had suffered from migraine for about twenty years. The attacks would come on him once or twice a week, sometimes lasting throughout the day, so that, though he often forced himself to carry on with his work, he was equally often compelled to give up and retire to bed in a darkened room. This meant that he could never be sure of fulfilling his obligations.

When he came to me, I treated him by needling acupuncture points liver 8 and a related point, gall bladder 20 (Fig. 23). The relationship of these two points illustrates, to the acupuncturist, the known physiological interaction between the liver and the gall bladder.

I treated the patient ten times at these two points. As a result, though he

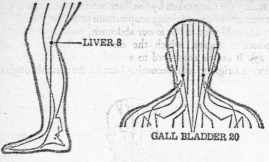

FIG. 23

still has an attack of the migraine about four times a year, he is otherwise well.

Branches of the Main Meridian

The main meridian has various subsidiary branches supplying areas of the body adjacent to it. The dotted line in Fig. 24 shows how the branch of the heart meridian traverses the lungs, goes to the big blood vessels entering and leaving the heart, penetrates the diaphragm and connects with the small intestine. Another part of this branch travels through the throat to the eye. It is not hard to see how the main meridian's sphere of influence is enlarged by its various branches.

Case History. An elderly woman came to see me some years ago complaining of her eyes, which were red, tender and sensitive to strong light. The stimulation of acupuncture point heart 3 at the elbow cured this condition, presumably because the upper branch of the heart meridian connects with the eye.

Indirect Course of the Main Meridian

Normally the course of the main meridian is taken as the line connecting various acupuncture points along the same meridian. The liver meridian, for example, runs over the inside of the leg and over the abdomen, thus influencing diseases along its course. It also influences not only diseases of the liver itself but those related to the liver by physiology, embryology, anatomy, function etc.

In reality the route taken by the liver meridian is much less direct. It changes its course by joining acupuncture points of other meridians above the ankle and in the lower abdomen, making a detour to the reproductive organs, which the direct course merely bypasses (Fig. 25). It can be compared to a traffic diversion on the original London to Brighton road becoming later on the main through route.

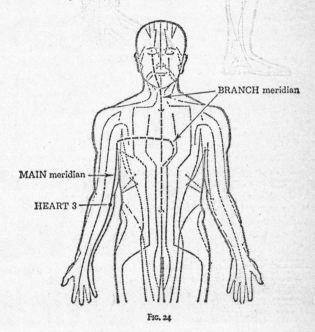

FIG. 24

Case History. A patient had painful periods, which caused her to remain in bed two days a month. Being a pharmacist by profession, she had tried various drugs unavailingly. I needled her once a month, halfway between the periods, at liver 8 on the inside of the knee till, after six treatments, she was cured and has been free of pain for the last ten years. From an acupuncture point of view one could say the cure was effected because the indirect course of the liver meridian goes to the reproductive organs.

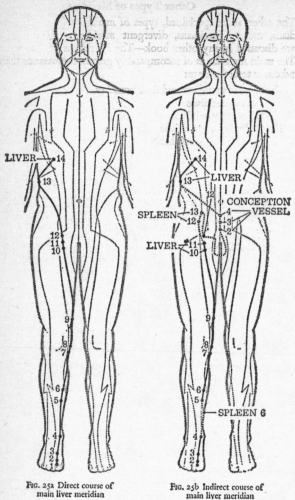

FIG. 25a Direct course of
main liver meridian

FIG. 25b Indirect course of
main liver meridian

Other Types of Meridians

The other, more specialised, types of meridians (connecting meridians, muscle meridians, divergent meridians, extra meridians) are discussed in my other book—The Meridians of Acupuncture. The main meridian is of incomparably greater importance than the others. In total there are:

The 12 main meridians and their branches.
The 8 extra meridians
The 12 muscle meridians
The 12 divergent meridians
The 15 connecting meridians

IV

THE ENERGY OF LIFE—QI*
including
THE EIGHT BODILY FORCES
OR SUBSTANCES

The ancient Chinese made no precise distinction between arteries, veins, lymphatics, nerves, tendons or meridians. They were concerned rather with a system of forces in the body, those forces which enable a man to move, to breathe, to digest his food, to think. As in other so-called primitive systems of medicine, like the Egyptian or the Aztec, the anatomical structures which make these physiological processes possible were not described in detail. They concentrated instead on this elaborate system of forces, whose interplay regulated all the functions of the body.

In Western medicine we have an intricate knowledge of anatomy, microscopic anatomy, the chemistry and biochemistry of the body, but little knowledge of what actually makes it 'tick'. It was this energy at the roots of all life which was the primary interest of the ancient Chinese.

LIFE ENERGY (QI)

Qi (life energy) is one of the fundamental concepts of Chinese thought. The manifestation of any invisible force, whether it be the growth of a plant, the movement of an arm or the deafening

*Pronounced chee as in cheese.

thunder of a storm, is called Qi: though, as we shall see, there are many varieties of it, each with its own specific function. In Hindu terminology the nearest equivalent to Qi is 'Prana'; in Theosophy and Anthroposophy it is called the 'Ether' or 'Etheric Body'.

Qi in the human body is called True Qi, and is created by breathing and eating. The Qi inhaled with the air is extracted by the lungs; the Qi in food and water by the stomach and its associated organ, the spleen.

> *True Qi is a combination of what is received from the heavens and the Qi of water and food. It permeates the whole body.'*
>
> (Ling Shu, cilie zhenxie pian)

Western medicine would explain death from asphyxiation as due to lack of air and death from starvation or dehydration to lack of food or water. The ancient Chinese neither ignored nor denied these obvious physical facts but they did not consider them the complete explanation. That must include the lack of the vital energy of life. The inability of the body to extract Qi from air, food and water is just as much a cause of death as its deprivation of them.

In these particular instances it is clear that the physical and metaphysical phenomena, deriving from the same source, cannot readily be distinguished. In others the difference can be more easily observed.

> *True Qi is the original Qi. Qi from Heaven is received through the nose and controlled by the wind-pipe; Qi from food and water enters the stomach and is controlled by the gullet. That which nourishes the unborn is the Qi of former heaven (pre-natal); that which fills the born is called the Qi of the latter heaven (post-natal)'.*
>
> (Zhangshi leijing)

Qi is universal:—

> *The root of the way of life (Dao or Tao), of birth and change is Qi; the myriad things of heaven and earth all obey this law. Thus Qi in the periphery envelops heaven and earth, Qi in the interior activates them. The source wherefrom the sun, moon and stars derive their light, the thunder, rain, wind and cloud their being, the four seasons and the myriad things their birth, growth, gathering and storing: all this is brought about by Qi. Man's possession of life is completely dependent upon this Qi.'*
>
> (Zhangshi leijing)

Before a mother can conceive and the foetus develop, her body, according to the Chinese, must be in harmony. The two extra meridians called the Vessel of Conception and the Penetrating Vessel should be active and the umbilical cord properly functioning. Only then can the Qi of former heaven be adequately received by the mother. After birth the Qi of Former Heaven is cut off, as the baby begins to take in from the air it breathes and from digestion of food and water the Qi of Latter Heaven.

Qi activates all the processes of the body, *'the unceasing circulation of the blood, the dissemination of fluid in skin and flesh, joints and bone-hollows, the lubrication of the digestive tract, sweating, urination, etc.'* In Chinese treatises on acupuncture the effect of pricking the skin with a needle is called 'obtaining Qi' and if the needle fails to obtain Qi (which is often indicated by various signs and symptoms) the acupuncture treatment will be ineffective.

'Thus one is able to smell only if Lung Qi penetrates to the nose; one can distinguish the five colours only if Liver Qi penetrates to the eyes; one can taste only if Heart Qi penetrates to the tongue; one can know whether one likes or dislikes food only if Spleen Qi penetrates to the mouth.'

'The capabilities of the seven holes (eyes, ears, nose and mouth) depend upon the penetration of the Qi from the five solid organs' (as mentioned in the preceding paragraph).

(Zhongyixue gailun)

'That which was from the beginning in heaven is Qi; on earth it becomes visible as form; Qi and form interact, giving birth to the myriad things.'

(Su Wen, tianyuanji dalun)

The cycle of changes which results from the interaction of Qi and form is what the Chinese meant when they described the 'transform-ation of air, food and water into Qi, blood and other substances. This is the transformation at work in the rhythms of growth and decay, in the changes from the flower to the fruit or the child to the old man.

The meridians are the tracks along which travel many of the impulses mentioned above—or, as the Chinese put it, *'the meridians are the paths of the transforming action of Qi in the solid and hollow organs.'*

(Yijiang jingyi)

The word Qi in Chinese has, besides 'Life Energy', the further meaning of 'Air'. Only over the last hundred years, since it became possible to weigh air by creating a true vacuum, has it been defined as physical. To the Chinese air was non-material and could therefore only be a vehicle for the forces of energy. Thus they often use the phrase 'bad Qi' for what we, more prosaically, would call a bad smell. This double meaning of air and energy may be an explanation of the breathing exercises in Indian Yoga; the Chinese use a similar system to obtain Qi.

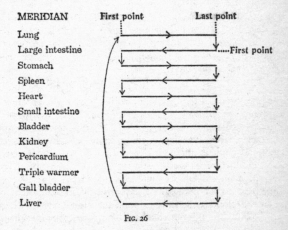

FIG. 26

The lungs are principally concerned with the Qi of the whole body. The spleen is concerned with the middle Qi, via its coupled organ the stomach which obtains the Qi from food. The kidney's Qi determines the hereditary constitution, as the production of semen and ova is largely determined by the kidney. (See section on the kidney in The Meridians of Acupuncture).

The meridian cycle begins with that of the lungs as Qi enters there. Also the Qi from the digestion of food and water in the stomach goes via the spleen to the lungs. Thus from the lungs, Qi of both sources is distributed around the body via the meridians in a certain order. Starting with the first point (or point of entry) of the

lung meridian, thence to the last point (or point of exit) of the lung
meridian, to the first point (or point of entry) of the large intestine
meridian etc. (Fig. 26).

NOURISHING QI (YING QI)

Qi is not stationary in the body; it circulates much as the blood has,
since the time of Harvey, been known to circulate. This circulation
is of two main types: that of Nourishing Qi, through the meridians
and blood-vessels, and that of Protecting Qi, between the skin and
the flesh in the subcutaneous tissues.

'*Man receives Qi in his food. Qi, entering the stomach, is transmitted
to the lungs, the five solid and the six hollow organs, so that all these may
receive Qi. The purer part of food is Nourishing Qi, the less pure part
Protecting Qi, Nourishing Qi being within the meridians and blood-vessels
and Protecting Qi outside them.*'

(Ling Shu, yingwei shenghui pian)

The purest part of the food digested in the stomach becomes the
Nourishing Qi, which circulates round the body following the
superficial circulation of energy, i.e. lungs, large intestines, stomach,
spleen, heart, small intestine, bladder, kidney, pericardium, triple
warmer, gall bladder and liver. But owing to their habit of drawing
no clear distinction between meridians, blood-vessels, lymphatics,
nerves and tendons, Chinese writers describe the Nourishing Qi
in some passages as flowing along the meridians, in others as accompanying the blood in its flow through the blood vessels, while in
others the blood itself is described as flowing along the meridians.

PROTECTING QI (WEI QI)

The Protecting Qi complements the Nourishing Qi and, like it, is
formed by the digestion of food and water in the stomach (and
spleen) and distributed thence to the rest of the body. But while the
Nourishing Qi is distilled from the purest elements, the Protecting
Qi emerges from the coarser products of digestion, and because of
this crude origin has rougher and more aggressive properties. It
therefore cannot penetrate the delicate meridians and vessels but
instead circulates in the subcutaneous tissues:—

*Wei Qi is the fierce Qi of food and water. Trembling, urgent, slippery,
sharp, it cannot enter the vessels and meridians but travels between the*

skin and flesh, vaporises in the diaphragm and scatters in the chest and abdomen.

(Su Wen, bi lun)

'*Wei comes out of a restless urgency of the fierce Qi and at first moves incessantly in the four limbs between the skin and flesh.*'

(Ling Shu, xieke pian)

The protecting Qi warms the subcutaneous tissues, moistens the skin, controls the opening and closing of the pores and nourishes the space between the skin and the flesh. But its most important function is the protection of the body from 'outside invading evils'. (See chapter X). If, for example, wind and cold invade the body, it meets the invasion by producing the desire for warmth and the manifestations of fever. Sweat is emitted, the fever subsides and the invading forces are dispersed. If, on the other hand, the invasion is successful, the patient will fall a victim to the disease.

When the Protecting Qi is too weak to permeate the subcutaneous tissues, the meridians will become empty and hollow, the flow of blood sluggish and uneven, the skin and flesh inadequately nourished.

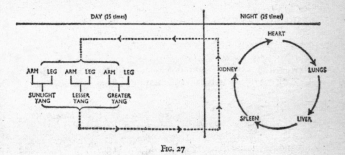

FIG. 27

The patient may then become a sufferer from rheumatism; or, if the wind cold and damp remaining in the body affect the meridians, vessels and joints, from an attack of arthritis.

The Nourishing Qi comes into the same category as the Yin, since it is composed of a rarefied substance and circulates with the blood in the interior of the body. The Protecting Qi could be classified under Yang, since it is composed of coarser elements, circulates in the surface of the body and is associated not with the blood-stream but with Qi. (See chapter V).

It is said that the Protecting Qi every 24 hours completes 50 cycles in the body, 25 cycles parallel with the Yang during the day and 25 parallel with the Yin during the night (Fig. 27). When it circulates through the Yang meridians in the daytime, it passes through those of the large intestine, the stomach, the triple warmer, the gall bladder, the small intestine and the bladder, in that order. Similarly at night, when it circulates through the Yin meridians, it passes from the kidney to the heart and thence to the lungs, the liver and finally the spleen.

THE BLOOD

'The middle warmer receives Qi, extracts the liquid and turns it red. This is called blood.'

(Ling Shu, juequi pian)

'Nourishing Qi collects fluid and pours it into the vessels, changing it into blood in order to nourish the four extremities and to flow into the solid and hollow organs.'

(Ling Shu, xieke pian)

Blood, then, is formed together with the Nourishing Qi from the process of digestion. To the digested food and water fluid is added, which becomes red in the middle warmer and then flows with the Nourishing Qi in its circuit of the meridians, blood vessels, organs, muscles and bones.

'Only if the vessels are so regulated that there is an uninterrupted circulation of blood can the skin, flesh, muscles, bones and joints of the body be strong, vigorous and supple. Thus the reason why the eyes can see, the feet walk, the hands grasp and the skin sweat is that they are all irrigated by blood.'

(Zhongyixue gailun)

Qi is thought to control the movement of the blood. 'Qi is the general of the blood; if Qi moves, then the blood moves.' Along the meridians and blood vessels blood and Nourishing Qi travel together.

'If the blood vessels and meridians are empty, so that the flesh cannot *obtain nutriment from blood and Nourishing Qi, symptoms of pain, itching and numbuess will result.'*

Or again:—

'If an evil invades the blood vessels, then the movement of the Nourishing Qi is hindered and the blood remains blocked within the flesh, causing symptoms of swelling.'

It is said that, in addition to Ying Qi, 'blood' also circulates in the meridians, the Ying Qi representing the Yang element, the 'blood' representing the Yin element. Each meridian has a certain proportion of Qi and 'blood' but the proportions differ, as may be seen from the following table:

SUNLIGHT YANG	More	Qi	More	GREATER YIN
	More	Blood	Less	
LESSER YANG	More	Qi	Less	ABSOLUTE YIN
	Less	Blood	More	
GREATER YANG	Less	Qi	More	LESSER YIN
	More	Blood	Less	

Heart – its principal function is blood
Spleen – gathers blood
Liver – stores blood

LIFE ESSENCE: ESSENCE AND SEMEN (JING)

The Chinese distinguished two types of the creative force (Jing) in the human body:—

(1) Jing—Essence

This type of the Life Essence is formed, (as is also the second type, the semen Jing) in the same way as the other bodily substances, by the transforming action of Qi on the food and water in the stomach and spleen. Thereafter it is stored in the kidneys. Whenever Life Essence is required, the kidneys inject it into the body so that it can circulate through the remaining organs. Thus, whenever the 'six evils' or 'seven emotions' (see chapter X) attack the body and injure it, they inevitably also injure this Life Essence.

(2) Jing—Semen

The semen type of Jing is present in both male and female, represented in the male by the spermatozoa, in the female by the ova:

'In the creation of man, appears first the Life Essence.'

(Ling Shu, jinjmo pian)

An embryo is formed by the union of male and female semen Jing. The essence resulting from this union, which nourishes the foetus, is known as the 'life essence of former heaven'. After birth, the child is nourished by the 'life essence of latter heaven' (derived from the food and water digested in the stomach) which also helps in the formation of Qi, Nourishing Qi, Protecting Qi, and blood.

According to the Chinese, the 'semen life essence' becomes mature in girls at the age of 14 (two periods of seven years), in boys at the age of 16 (two eight-year periods). Likewise the menopause occurs in women after seven seven-year periods at the age of 49, while the equivalent happens in man after eight eight-year periods at 64. Thus, ideally, a woman is able to reproduce between the ages of 14 and 49, a man between 16 and 64.

SPIRIT (SHEN)

Shen is usually translated as 'spirit', a word which to the Western mind more often than not suggests the supernatural. But Shen, in common with most other concepts in traditional Chinese thought, is a down-to-earth word. The spirit is created by the normal processes of reproduction: it needs man's heart as a house to live in; it must have food and water to nourish it and, if it is unable to function properly, the result will be actual physical disease.

'The origin of Life is in the Life Essence (the male and female semen). When these two unite to make one, that is called the Spirit.'

(Ling Shu, benshen pian)

'Basically the Spirit is the Life Essence Qi derived from food and water.'
(Ling Shu, pingren juegu)

'The five tastes enter the mouth and are stored in the stomach and intestines, in order to nourish the five Qi. The five Qi then unite creating

fluid, and the Spirit is then formed as a natural consequence of this process.
(Su Wen, liujie zangxiang lun)

Born from the union of male and female semen, sustained in its earthly existence by food and water, the Spirit dwells in the heart, while other human attributes inhabit other organs:—

'*The heart houses the Spirit, the lungs the Animal Soul, the liver the Spiritual Soul, the spleen the Mind and the kidneys the Will.*'
(Su Wen, xuanming wuqi lun)

It is for this reason that a cardiac disease affects the Spirit to a greater degree than one elsewhere in the body. The physical functions of the Spirit are well documented in Chinese literature:—

'*At a hundred years of age the five solid organs are empty, the Spirit and the Qi have gone completely and form alone exists; that is all.*'
(Ling Shu, tiannian pian)

'*Whoever has the Spirit, flourishes; whoever loses the Spirit, perishes.*'
(Su Wen, yijing bianqi lun)

Or (to put it as the Chinese might have done) the seeing of the eyes, the hearing of the ears, the speaking of the mouth, the movement of the limbs and body, the consciousness and activity of the mind, all are manifestations of the Spirit informing the flesh. If the Spirit is weak, the eyes are dull, the vitality exhausted; in severe cases the speech may become abnormal and the mind so affected as to cause hallucinations and delusions, agitation, delirium and unconsciousness.

FLUID

The amount of Fluid in the body chiefly depends on the amount of food and water digested. The liquid in the stomach is metabolised by the action of the Yin Fluid already there and, as a result, takes on a special quality which differentiates it from water outside the body. It is living water; it has 'acquired Life Essence'.

This Chinese doctrine that Fluid is, as it were, a living individual entity means that the amount of Fluid in the body is as important as its quality. Too much or too little upsets the Yin-Yang balance within the system. So Fluid is regulated by various organs. The lungs control the process of energising by the action of Qi, the kidneys the amount of water to be used or rejected, the bladder stores the Fluid and the triple warmer manages the drainage. It is said, too, that the small intestine divides the liquid into 'pure' and 'impure', the 'pure' becoming Fluid and being distributed for the nourishment of the rest of the body, the 'impure' passing into the bladder to be excreted. The seasons also influence the amount of Fluid in the body, for perspiration increases in the hot weather and urination in the cold. Any failure in this system of control and response can lead to illness and it is said that the oedema and diabetes are both diseases of Fluid.

The Chinese distinguish two types of Fluid, the Clear Fluid (Jin), which is of the nature of Yang and circulates with the Protecting Qi, and the Thick Fluid (Ye), which is Yin and circulates with the Nourishing Qi and the blood in the blood vessels.

'*When the pores leak and sweat is emitted, it is called Clear Fluid.*'

'*Thick Fluid is poured into the bones, so that they have the property of bending and stretching; if broken, they leak Thick Fluid.*'

(Ling Shu, juenqi pian)

The Clear Fluid, then, moistens the flesh and skin and exudes as normal sweat; the Thick Fluid keeps the sinews pliant, lubricates the joints, fills the marrow of the bones and the hollows of the brain and exudes on to the surface of the body as the greasy excretion of the sweat glands.

By the action of the middle warmer Fluid and Nourishing Qi become blood and each of the five solid organs likewise has a transforming effect on the Fluid. The liver transforms it to tears, the heart to sweat, the spleen to saliva, the lungs to mucus, the kidneys to urine.

CORRELATIONS OF THE EIGHT BODILY HUMOURS

In harmony with the general nature of Chinese thought, the eight bodily humours are seen as interdependent, not isolated: some of them wax while others are waning, some wax or wane together. The inter-related effects vary according to circumstances.

Qi is the great energiser. It causes the blood to circulate and the Fluid to be disseminated throughout the body and excreted as urine and sweat. It transforms the food we eat into Life Essence, blood, Fluid, etc. and the organs of the body are active only because these elements which transfuse them have been energised by Qi. It activates the process of digestion and regulates the capacity of the intestines to absorb only those substances which the body needs and to excrete the remainder.

In Chinese texts it is often not possible to distinguish between blood and Nourishing Qi, for they both flow along various pathways (blood vessels, meridians, etc.) to nourish the body and, if the flow of the one is interrupted so is that of the other, with the result in each case of symptoms of 'pain, itching and numbness'.

Blood and Nourishing Qi are the basic constitutents of what the Chinese call Jingshen, a combination of Life Essence and Spirit which might be translated as 'Vitality' in both its mental and physical aspects. This is an example of a Vitality disease caused by an excessive loss of blood after childbirth:—

'When a woman has given birth, fluid has been lost, the blood is empty and the mind weak, causing her Vitality to become confused, her speech delirious etc., and, in extreme cases, resulting in madness. This may be treated by tonifying the blood and stimulating the heart, pacifying the heart and stimulating the mind.'

(Guaiji zonglu·chanhoumen)

A man whose Vitality is excessive will be prone to violent rages, sometimes so violent that, as the blood and Qi surge upwards, he will actually vomit blood. Moreover, if the Spirit is not at ease then the heart in which it dwells will be uneasy also and the blood and Nourishing Qi weak.

In comparing blood and Nourishing Qi with Qi and Protecting Qi, the Chinese held that the first two nourish the interior of the body, while the second two protect its surface. Each of these pairs assists the other; for nourishment cannot proceed if the surface of the body is unprotected and, without nourishnent, the forces protecting the surface cannot do so. It is not wise, in other words, to fight a war on two flanks. Although in this context it is the function of the Qi to protect the exterior and of the blood to nourish the interior of the body, in another sense the functions are not separate but interchanged; for the blood is also nourishing the skin on the surface while the Qi energises the organs within.

'Yin in the interior is the guardian of Yang; Yang in the exterior motivates Yin.'

(Su Wen, yinyang yingxiang dalun)

It is said that blood is created by Qi and that blood is the basis of Qi.

'*If Qi and blood are not evenly balanced, then Yin and Yang will oppose one another; Qi will rebel against Protecting Qi, blood against Nourishing Qi, blood and Qi will be separated, one being full and the other empty.*'

(Su Wen, tiaojing lun)

To the ancient Chinese, Life Essence Qi and Spirit were 'the three precious things', the basis of all being; for Life Essence is formed from food and water, Qi from food, water and air, and both Qi and Life Essence are combined in Spirit. The Spirit, they said, flourishes in one whose Life Essence and Qi are sufficient; if it does not flourish, then the Life Essence is weak.

'*Although the Spirit is produced from Life Essence and Qi, nevertheless that which governs and selects Life Essence and Qi and controls their function, is the Spirit of the heart.*'

(Zhangshi leijing)

Because of their common origin, a deficiency of Fluid entails a corresponding deficiency of Qi and blood, and vice versa. A patient suffering from excessive sweating, vomiting and diarrhoea, with a resultant loss of Fluid, will also have what the Chinese call 'Qi and blood deficiency symptoms'—shortness of breath, shallow breathing, fine pulse, palpitations and coldness of the limbs. Conversely, a loss of blood will produce 'Fluid deficiency symptoms'—dryness of the mouth, constipation and infrequent micturition. Sweat being a transformation of Fluid, a patient who has lost blood or is otherwise dehydrated cannot easily perspire.

V

THE PRINCIPLE OF OPPOSITES

The Chinese believed that in the beginning the world was a formless indivisible whole. There was no distinction between heaven and hell, fire and water, day and night; there was neither birth nor death, growth nor decay; all imaginable things were merged together without definition in an unchanging unity. Had man existed, he would have remained forever incapable of evolution, a static and perfect image.

For life as we know it to be possible with all its richness and variety, its infinite potentialities for good and ill, this world had to be split in two. The Unity had to become a duality; and from this duality arose the idea of the complementary opposites, the negative and the positive, which the Chinese called the Yin and the Yang. These two principles are at the very root of the Chinese way of life; they pervade all their art, literature and philosophy and are therefore also embodied in their theories of traditional medicine.

These principles are of course, up to a point, accepted in the West. We too divide every phenomenon into its two contrary components. Male and female, hard and soft, good and bad, positive and negative electrical charges, laevorotary and dextrorotary chemical compounds —all these are 'opposites'. It is indeed a fact that nothing can happen in the physical world unaccompanied by positive or negative electrical changes. If a man moves his hand or a raindrop falls or a child rolls a marble across the floor, such changes will affect the balance of positive and negative charges in each of these instances. But in Europe we have not formulated this polarity as a universal law as have the Chinese, to whom the perpetual interplay of the Yin and the Yang is the very keystone of their thinking. It is the law operating throughout all existence that the states of Yin and Yang must succeed one another, so that, in a Yin condition, the corresponding Yang state can be precisely foretold. The practical application of this law to acupuncture can be illustrated thus:—

	Yang	Yin
In the natural world:	Day	Night
	Clear day	Cloudy day
	Spring/Summer	Autumn/Winter
	East/South	West/North
	Upper	Lower
	Exterior	Interior
	Hot	Cold
	Fire	Water
	Light	Dark
	Sun	Moon
In the body:	Surfaces of the body	Interior of body
	Spine/back	Chest/abdomen
	Male	Female
	Clear or clean body fluid	Cloudy or dirty body fluid
	Energy (Qi)	Blood
	Protecting Qi	Nourishing Qi
In disease:	Acute/virulent	Chronic/non-active
	Powerful/flourishing	Weak/decaying
	Patient feels hot or hot to touch or has temperature	Patient feels cold or cold to touch or has under-temperature
	Dry	Moist
	Advancing	Retiring
	Hasty	Lingering

The twelve basic organs and meridians are similarly divided into the Yang hollow (Fu) organs, which 'transform but do not retain' and the Yin solid (Zang) organs, which 'store but do not transmit':—

Yin	Yang
Liver	Gall bladder
Heart	Small intestine
Spleen (Pancreas)	Stomach
Lung	Large intestine
Kidney	Bladder
Pericardium	Triple warmer

The qualities of Yin and Yang are relative, not absolute. For example, the surface of the body is Yang, the interior Yin. But this relation also remains constant within the body, for the surface of every internal organ is always Yang and its interior always Yin, down to the individual cells that compose it. Similarly, a gas is Yang, a solid Yin; but among the gases the more rarefied are Yang, the denser are Yin. Life and death belong to Yang, growth and storage to Yin, so that 'if only Yang exists, there will be no birth: if only Yin exists, there will be no growth.' The life of every organism depends upon the correct balance of its various components.

'Yin and Yang are the Tao of heaven and earth (the basic law of opposition and unity in the natural world), the fundamental principle of the myriad things (all things can only obey this law and cannot transgress it), the originators (literally parents) of change (change in all things is according to this law), the beginning of birth and death (the birth and creation, death and destruction of all things begins with this law), the storehouse of Shen Ming (the location of all that is mysterious in the natural world). The treatment of disease must be sought for in this basic law (man is one of the living things of nature, so the curing of disease must be sought for in this basic law).'

(Su Wen, yinyang yingxiang dalun)

Since everything in life can be classified according to its Yin and Yang components, it is said:—

'Now the Yin/Yang has a name but no form. Thus it can be extended from one to ten, from ten to a hundred, from a hundred to a thousand, from a thousand to ten thousand (i.e. it can embrace all things).'

(Ling Shu, yingang xi riyue pian)

Each component not only opposes but also contains its opposite, for:-

'There is Yin within the Yin and Yang within the Yang. From dawn till noon the Yang of Heaven is the Yang within the Yang; from noon till *dusk the Yang of heaven is the Yin within the Yang; from dusk till midnight the Yin of heaven is the Yin within the Yin; from midnight till dawn the Yin of heaven is the Yang within the Yin.'*

(Su Wen, jinkui zhenyan lun)

Thus 'functional movement' belongs to Yang, 'nourishing substance' to Yin, nor can the one exist without the other; for, if the intestines and other internal organs do not move, 'nourishing substances' cannot be digested and, if over a long period 'nourishing substances' are not provided, the organs cease to move.

'Yin in the interior is the guardian of Yang; Yang in the exterior is the activator of Yin.'

(Su Wen, yinyang yingxiang dalun)

The opposition of Yin and Yang is not static; it is a perpetually changing rhythm of movement, whose interplay produces growth, transformation and death.

'The relation of Yin and Yang is the means whereby the myriad things are able to come to birth, Yin and Yang react upon each other, producing change.'

(Su Wen, yinyang yingxiang dalun)

This changing rhythm in the balance of Yin and Yang ensures that there is never an excess of either of these polar opposites, for over-activity of Yang is at once adjusted by the yielding passivity of Yin.

'In Winter on the 45th day (the beginning of spring) the Yang Qi is slightly superior and the Yin Qi slightly inferior; in summer on the 45th day (the beginning of autumn) the Yin Qi is slightly superior, the Yang Qi slightly inferior.'

(Su Wen, maiyao jingwei lun)

In the former case, Yang Qi waxed with the upsurge of spring as Yin Qi waned; in the latter, Yin Qi waxed with the decline to winter as Yang Qi waned.

'When speaking of Yin and Yang, the exterior is Yang, the interior is Yin; when speaking of Yin and Yang in the human body, the back is Yang, the abdomen Yin; when speaking of Yin and Yang of the Zang and Fu in the body, then the Zang are Yin, the Fu are Yang; liver, heart, spleen, lungs and kidney are all Yin, the gall bladder, stomach, large intestine, small intestine, bladder and triple warmer are all Yang.'

'*Thus the back is Yang and the Yang within the Yang is the heart.
The back is Yang and the Yin within the Yang is the lungs.
The abdomen is Yin and the Yin within the Yin is the kidneys,
The abdomen is Yin and the Yang within the Yin is the liver.
The abdomen is Yin and the extreme Yin within the Yin is the spleen.*'

(Su Wen, jinkui zhenyan lun)

If this balance of Yin and Yang is upset there is a reaction:—

'*Excess of Yin causes a Yang disease, excess of Yang a Yin disease.
Yang in excess produces heat and, if the heat is extreme, it will produce
cold; Yin in excess produces cold and, if the cold is extreme, it will produce
heat.*'

(Su Wen, yinyang yingxiang dalun)

In diagnosis:

'*The skilled diagnostician examines the countenance and feels the pulse.
First dividing them into Yin and Yang, he judges the pure (Yang) and the
impure (Yin) and thus knows the diseased part of the body... He feels
the pulse to ascertain whether it is floating (Yang), deep (Yin), slippery
(Yang) or rough (Yin) and knows where the disease originated. If there is
no mistake in his diagnosis then nothing 'will be overlooked.*'

(Su Wen, yinyang yingxiang dalun)

In the treatment of disease, if Yang is hot and over-abundant,
thus injuring the Yin fluid (Yang excess causing a Yin disease),
the surplus Yang can be decreased by a method called 'cooling what
is hot'. If Yin is cold and over-abundant, thus injuring the Yang
Qi (Yin excess causing Yang disease), the surplus Yin can be decreased by the method called 'heating what is cold'. Conversely,
if Yin fluid is deficient and so, unable to control the Yang, causes
it to become violent; or if Yang Qi is deficient and, unable to control
Yin, causes it to become over-abundant, then the deficiency must be
tonified. The Neijing describes the method thus:— '*In Yang diseases
treat the Yin; in Yin diseases treat the Yang.*'

Ever since the days of the legendary Yellow Emperor, preventive
medicine has to some extent depended on keeping the activities of
the body in harmony with the rhythm of the changing seasons.
Therefore, as the Su Wen shaggutian zhenlun puts it: '*Harmonise
with the Yin and Yang, harmonise with the four seasons.*' And again:—

'*The Yin and the Yang of the four seasons are the basis of the myriad things. Therefore a wise man will nourish Yang in spring and summer, Yin in autumn and winter. Follow this fundamental law and you will be on the threshold of birth and growth; rebel against it and you will destroy its root and harm its truth. For Yin and Yang and the four seasons are the beginning and end of the myriad things, the roots of life and death. If you rebel against them you will destroy life, if you follow them disease will not arise... He who follows Yin and Yang will have life; he who rebels against them will die. Obey, and you will be cured; rebel and calamity will follow.'*

(Su Wen, siqi tiaoashen dalun)

The Chinese held that the causes of disease come not only from outside, as, in the above quotation, from a disharmony with the seasons, but also from within the body itself. Foremost among these causes of disease are the emotions:—

'Violent anger injures Yin; violent joy injures Yang.'

(Su Wen, yinyang yingxiang dalun)

Man should live in harmony with heaven and earth for, while his feet rest always on the earth, his mind can reach upwards beyond the remotest stars.

'If you understand, above, the writings of heaven (astronomy): below, the principles of earth (geography): and, in between earth and heaven, the affairs of man: then may you have long life.'

(Zhu zhijiao, pian)

'If in curing the sick you do not observe the records of heaven nor use the principles of earth, the result will be calamity.'

(Yin yang, yingxiang dalun)

This interplay of heaven and earth was thought to be the beginning of all life. Before that there was the Great Void, which nothing created, nothing preceded, nothing sustained, till it was brought into movement by the Great Qi of the universe. Then the Qi of heaven began to descend and the Qi of earth to ascend and from their interplay came change, movement and transformation; and thus there was life. With the beginning of movement came the beginning of silence, the rhythms of activity and quiescence producing more life, further changes.

The intercourse of Qi between heaven and earth resulted in the creation of man.

'*In heaven there is wind, in earth there is wood; in heaven there is heat, in earth there is fire; in heaven there is damp, in earth there is earthiness; in heaven there is dryness, in earth there is metal; in heaven there is cold, in earth there is water; in heaven there is Qi, in earth there is form; form and Qi interact, thus creating the myriad things.*'

(Su Wen, tian yuanji dalun)

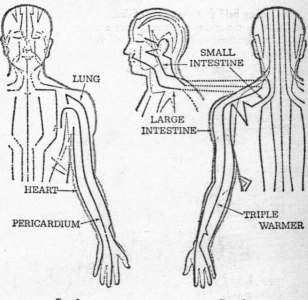

FIG. 28a   FIG. 28b

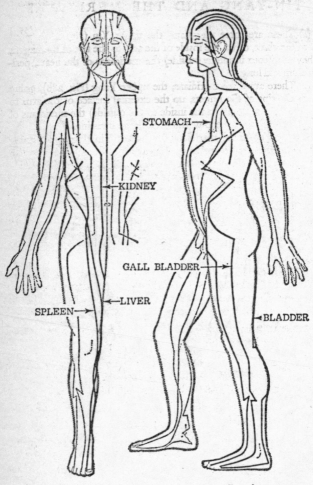

STOMACH→

←KIDNEY

GALL BLADDER—→

SPLEEN→   ←LIVER   ←BLADDER

Fig. 29a   Fig. 29b

YIN-YANG AND THE MERIDIANS

1 (a) There are three meridians, the upper Yin (Fig. 28a), going from the chest, down the inside of the arm to the tips of the fingers. They are (from inside to outside) the meridians of the heart, pericardium and lung.

(b) There are three meridians, the upper Yang (Fig. 28b), going from the tips of the fingers, up the external surface of the arm to the face. They are (from the inside to the outside) the meridians of the small intestine, triple-warmer and large intestine.

(c) There are three meridians, the lower Yin (Fig. 29a), going from the toes, up the inside of the legs, over the abdomen to end on the front of the chest, near the origin of the upper Yin. They are (from the inside to the outside—in adult anatomy) the meridians of the spleen, liver and kidney.

(d) There are three meridians, the lower Yang (Fig. 29b) going from the head, down the body and the external surface of the legs to the toes. They are (from the inside to the outside—in adult anatomy) the meridians of the stomach, gall bladder and bladder.

2 The Chinese normally speak of the meridians in pairs and they distinguish the members of each pair by reference to the arm or leg, thus indicating the main location of the particular meridian instead of the particular organ to which it is related:

Sunlight Yang	arm—large intestine leg—stomach
Lesser Yang	arm—triple warmer leg—gall bladder
Greater Yang	arm—small intestine leg—bladder
Greater Yin	arm—lung leg—spleen
Absolute Yin	arm—pericardium leg—liver
Lesser Yin	arm—heart leg—kidney

Thus the small intestine meridian is referred to by the Chinese as the Arm Greater Yang, and the bladder meridian as the Leg Greater Yang. Used on its own the term Greater Yang would refer to the two meridians jointly.

3 Other interrelationships are shown in Fig. 30. (see also chapter VII and The Meridians of Acupuncture).

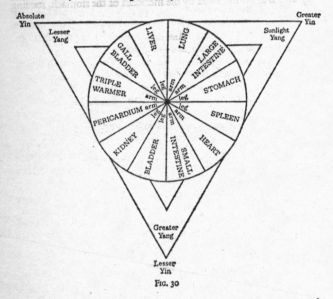

FIG. 30

4 If the arms and legs are held in the normal walking position with the thumb and toes anteriorly whilst the little finger and heel are posterior, the following arrangement ensues, whether it be the arm or leg (Fig. 31).

5 If the arms and legs are held in the embryological position with the medial surface of the arm and leg turned anteriorly, the positions of the twelve meridians are more easily followed.

FIRST MERIDIAN CYCLE (Fig. 32).

(*a*) The meridian of the lung starts on the chest and then runs down the *outer* side of the *anterior* surface of the arm to the thumb.

(*b*) This meridian is followed by the meridian of the large intestine, that starts at the end of the index finger and runs up the *outer* side of the *posterior* surface of the arm to end at the nose.

(*c*) This is again followed by the meridian of the stomach, starting

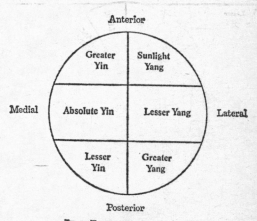

FIG. 31 Transverse section of arm or leg

below the eye and running down the *outer* side of the head (relative to bladder and gall bladder meridians), chest and abdomen and the *outside* of the external (embryologically *posterior*) surface of the leg, to end at the end of the second toe.

(*d*) Finally, the meridian of the spleen starts at the big toe, which, although it is apparently the inside of the foot, is in fact embryologically the *outer* side of the foot. This meridian runs up the *outside* of the medial (embryologically *anterior*) side of the leg and the outside of the *anterior* surface of the abdomen and chest to end near the origin of this first meridianal cycle, not far from the starting point of the meridian of the lung.

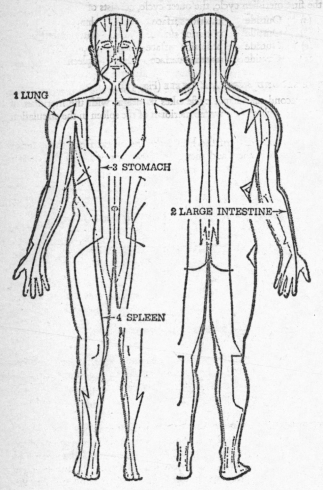

FIG. 32

Thus following the embryological surfaces of the human body, the first meridian cycle, the outer cycle, consists of:

(a)	Outside	anterior surface	arm	lung
(b)	Outside	posterior surface	arm	large intestine
(c)	Outside	posterior surface	leg	stomach
(d)	Outside	anterior surface	leg	spleen

THE SECOND MERIDIAN CYCLE (Fig. 33).

The second meridianal cycle takes its origin with the meridian of the heart, which follows the meridian of the spleen in the circulation of energy.

(a) The meridian of the heart takes its origin from the anterior surface of the chest and then runs down the *inside* of the *anterior* surface of the arm to end at the little finger.

(b) This is followed by the meridian of the small intestine which takes its origin at the end of the little finger and runs up the *inside* of the *posterior* surface of the arm to the cheek.

(c) This is followed by the meridian of the bladder that starts at the nose, runs over the *inside* of the skull and down the *inside* of the *posterior* surface of the back and leg to end at the little toe, which embryologically is on the *inside*.

(d) Finally, the meridian of the kidney starts on the sole of the foot and runs up the *inside* of the embryologically *anterior* surface of the leg, abdomen and chest, to end the second meridianal cycle near its origin on the *anterior* surface of the chest.

The second meridianal cycle, the inner cycle, is thus made up as follows:

(a)	Inside	anterior surface	arm	heart
(b)	Inside	posterior surface	arm	small intestine
(c)	Inside	posterior surface	leg	bladder
(d)	Inside	anterior surface	leg	kidney

THE THIRD MERIDIAN CYCLE (Fig. 34).

(a) The third and final meridianal cycle, starts as all others on the front of the chest, with the meridian that follows the kidney meridian in the circulation of energy, namely the meridian of the pericardium. This runs down the *middle* of the *anterior* surface of the arm, be-

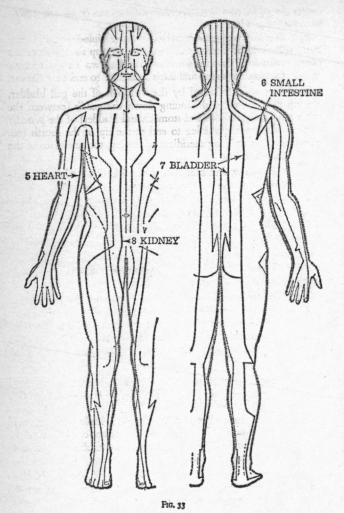

FIG. 33

tween the other two Yin meridians of the arm (heart and lung), to end at the middle finger.

(b) This is followed by the meridian of the triple-warmer, which starts at the end of the fourth finger and runs up the *middle* of the *posterior* surface of the arm between the other two Yang meridians of the arm (small intestine and large intestine), to end near the ear.

(c) This is again followed by the meridian of the gall bladder, starting near the ear and running down the *middle* (between the other two Yang meridians of stomach and bladder) of the *posterior* surface of the trunk and leg to end at the tip of the fourth toe, between the other Yang meridians of the leg, the meridians of the stomach and bladder.

(d) Finally the twelfth meridian, that of the liver, takes its origin on the lateral side of the big toe, between the origins of the other Yin meridians of the leg, those of the spleen and kidney. Thereafter the liver meridian goes up the *middle* of the medial (embryologically *anterior*) surface of the leg, over the abdomen to end at the lower end of the chest, near the origin of the first meridian cycle.

The order of the 3rd, middle, meridianal cycle is thus:

(a)	Middle	anterior surface	arm	pericardium
(b)	Middle	posterior surface	arm	triple warmer
(c)	Middle	posterior surface	leg	gall bladder
(d)	Middle	anterior surface	leg	liver

These relationships of the outer, inner and middle meridians may be best visualised by regarding each group in a major part of the body:

(a) With the upper Yin, (lung, pericardium, heart, from outside to inside) on the anterior surface of the chest and arm, the relationship is obvious.

(b) With the upper Yang, (large intestine, triple warmer, small intestine, from outside to inside) on the posterior surface of the arm, the relationship is again obvious. The complicated course on the head is discussed under (e).

(c) The lower Yang meridians (stomach, gall bladder, bladder, from outside to inside) follow a straightforward course on the neck, trunk and legs if account is taken of:

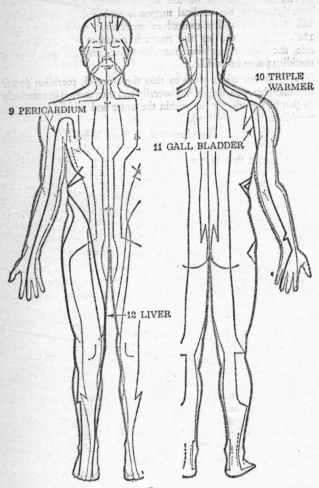

9 PERICARDIUM

10 TRIPLE WARMER

11 GALL BLADDER

12 LIVER

FIG. 34

(i) The lateral surface of the leg is embryologically speaking posterior, while the medical surface is embryologically anterior; and that the foot has twisted in embryonic evolution through 180°. Then on the leg the stomach meridian is posterior-outside, the bladder meridian posterior-inside, and the gall bladder meridian posterior-middle.

(ii) A shift has taken place so that the stomach meridian (outer meridian) has moved right out laterally and then moved in anteriorly so that in the adult it lies within the outer and middle lower Yin meridians (spleen and liver).

(d) The lower Yin meridians (spleen, liver, kidney, from outside to inside) follow the embryological pattern along their whole course except that in the lower leg the liver meridian (middle meridian) moves anterior (embryologically laterally) to the outer position. The kidney meridian takes its origin from the sole of the foot near the base of the third toe which is in agreement with the embryological condition.

(e) On the head the arrangement of the meridians is more complicated:-

The lower Yang meridians (stomach, gall bladder, bladder) follow the correct order: the bladder meridian on the inside, the gall bladder meridian in the middle, and the stomach meridian on the outside, if account is taken of the anterior shift of the stomach meridian as mentioned under (c).

The upper Yang meridians (large intestine, triple warmer, small intestine) follow the correct order except on the head and neck, where the large intestine meridian has moved two meridians anteriorly. This is similar to the anterior movement of the stomach meridian with which the large intestine meridian is coupled as the 'Sunlight Yang'.

There are still some obvious inconsistencies in the course of the adult meridians and the true embryological order.

The above theory may be coincidence, with as little meaning as some of the traditional Chinese ideas, for acupuncture abounds with theories involving numerical relationships, and anyone with a mathematical mind can easily invent new ones.

6 The anterior-posterior, or Yin-Yang relation of the meridians produces coupled meridians. The Chinese use the term outside and inside meridians (instead of coupled meridians) in reference to say the outer part and the lining of a coat, or in this context to the outer and inner part of the limbs.

Anterior or Yin	Lung	} Coupled meridians
Posterior or Yang	Large intestine	} or organs
" "	Stomach	}
Anterior or Yin	Spleen	} "
" "	Heart	}
Posterior or Yang	Small intestine	} "
" "	Bladder	}
Anterior or Yin	Kidney	} "
" "	Pericardium	}
Posterior or Yang	Triple warmer	} "
" "	Gall bladder	}
Anterior or Yin	Liver	} "

Each of the coupled meridians belongs to one of the five elements as discussed in the next chapter.

VI

THE FIVE ELEMENTS

'The five elements: wood, fire, earth, metal, water, encompass all the phenomena of nature. It is a symbolism that applies itself equally to man.'
(Su Wen)

The Chinese divided the world into five elements and everything on the earth was considered to belong, by its nature, to one or several of these five categories. These will recall to mind the

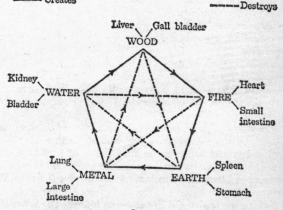

FIG. 35

four elements which were a familiar part of western practice up to recent times. These were: earth, water, air, fire; nor is it difficult to see (at least for anyone acquainted with this type of thought) that everything in the world belongs of its essence to one or several of these categories. For example, a brick belongs to the element earth; a glass of wine to the elements earth (glass) and water (wine); a barrage balloon to the elements earth (the balloon) and air (helium), a coal fire to earth (coal), air (carbon dioxide and other gases) and fire.

The above four elements are common to both the European and the Chinese systems. (Air = Metal). The fifth element designated 'wood', is only known to the Chinese and certain other civilizations whose roots extend to prehistoric times.

As mentioned, the five elements are:

> Wood
>
> Fire
>
> Earth
>
> Metal (Air in the western tradition)
>
> Water

The macrocosm, of which the human constitutes the microcosm, is considered the resultant of the interplay of these five primeval forces, which are linked in an unvarying pattern, one to another, in the manner illustrated in the chart (Fig. 35).

The outer lines represent the creative and the inner ones the destructive forces:

Wood will burn to create a *Fire*, which, when it has finished burning, leaves behind the ashes, *Earth*; from which may be mined the *Metals*; which, if heated, become molten like *Water*; which is necessary for the growth of plants and *Wood*. This is the creative cycle.

Wood destroys *Earth*, i.e. plants can crack rocks and break up the soil. *Earth* destroys *Water*, i.e. a jug with its earthenware sides prevents water from following its natural law of spreading out. *Water* destroys *Fire*, i.e. water, poured over a fire, will extinguish it. *Fire* destroys *Metal*, i.e. by melting. *Metal* destroys *Wood*, i.e. by cutting.

In medicine the law of the five elements is applied as follows:

		Yin	*Yang*
Wood is equivalent to the		Liver	and Gall bladder
Fire	„	„ Heart	and Small intestine
Earth	„	„ Spleen	and Stomach
Metal	„	„ Lung	and Large intestine
Water	„	„ Kidney	and Bladder
Fire	„	„ Pericardium	and Triple warmer

It should be understood that the Chinese when they used the terms 'Wood', 'Fire', etc., did not use the words in the actual restrictive sense of the physical wood, fire, etc., but rather as implying an archetypal idea in the sense in which it is used by the psychologist Jung, who studied Chinese philosophy profoundly. For example, the *idea* of the genus house is opposed to the idea of an *actual* house. Before it is possible to build a house it is necessary to have conceived the idea of 'house', whether this be a bungalow, a skyscraper, a modern glass and concrete affair, or an imitation Tudor perpetration. The general generic idea of 'house' is primary and covers a vast number of possibilities; an actual individual physical house made of bricks, etc. is only secondary to the general idea comprising all houses.

Thus what I have expressed above as '*Metal* destroys *Wood*, i.e., wood may be cut by a metallic saw;' is really a material vulgarisation of what is essentially an idea, an idea which may manifest itself in various physical guises such as wood, or the liver.

In the actual practice of acupuncture this theory of the five elements dictates that when the liver (wood) is tonified, the heart (fire) will automatically be also tonified, while the spleen (earth) is sedated; or, if the kidney (water) is sedated, the liver (wood) will also automatically be sedated, while the heart (fire) will be tonified.

The effect is similar with the related Yang organs. e.g. Tonification of the gall bladder will produce tonification of the small intestine and sedation of the stomach.

This law may seem to Western minds like the fanciful application of a philosophical law. Nevertheless, it operates whether one wishes it or not, provided the conditions of its working are complied with. For example, if the liver and heart are underactive and the spleen is overactive, tonification of only the liver will produce equilibrium between all three. If the heart were overactive or the spleen underactive the conditions for the operation of the law would not be met and there would be no result except for toni-fication of the liver.

Certain of these relationships are obvious in the practice of ordinary medicine, i.e., tonification of the kidney (water) will pro-duce by its increased excretion of water and solids a tonification (decongestion) of the liver (wood) and also a sedation of the heart (fire) which no longer has to force too much fluid through the body. A fuller explanation, however, will require much more research from the side of ordinary science.

The interplay of the creative and destructive forces in the five elements is, to the Chinese, another aspect of that delicate balance of all life in the polarity of Yin and Yang, which has been mentioned earlier. Only when this balance is upset and an organ cannot correctly react to a stimulus will disease result. .

If, for example, the heart were weak and nothing else could be done but to tonify it, the consequences in the body would resemble the pile-up of a traffic jam and the patient might fall seriously ill. For to tonify only the heart would at once increase its pumping action, and therefore also the circulation of the blood. But, in order to pump more vigorously, the muscular wall of the heart requires more oxygen: so the breathing becomes deeper or more rapid. Greater activity of the heart causes the release of more metabolites, which (since these have to be excreted) increases the action of the kidneys. Moreover, since the cardiac muscle needs more glycogen to fuel its energy, the liver too must come into action to release this.

In physics, chemistry, mechanics and all the sciences concerned with the non-living, the Western mind is accustomed to express such interactions in terms of exact laws. We accept, for example, un-questioningly the second law of thermodynamics, that every action has an opposite and equal reaction; or the law that the momentum of an object is proportional to its mass and the square of its velocity. But in the biological sciences, among which is that of medicine, exact laws are, to the Western mind, virtually non-existent. The Chinese, however, aimed at introducing into biology the same precision, at least in a qualitative if not a quantitative manner, as we have into mechanics. The paragraph above shows how the tonification of the heart cannot be considered as an act in itself, unrelated to the many accompanying readjustments it entails. In the West these are seen as physiological events; the Chinese expressed them in laws of almost mathematical precision.

· In certain abnormal conditions the balance of the five elements is not maintained. Normally the tonification of the liver sedates the spleen; but sometimes the reverse can happen and the tonification of the spleen will sedate the liver. This process is called 'mutual aid' and 'mutual antagonist'.

'If there is a surplus of Qi, then control that which is already winning and antagonise that which is not winning; if there is a deficiency of Qi, then antagonise and regulate that which is not winning, and bring out and antagonise that which is winning.'

(Su Wen, wuyunxing dalun)

This shows how an excess or deficiency of the five elements can play havoc with the laws which govern normal conditions. For example, a surplus of water Qi may destroy fire Qi (that which is winning) but may also insult earth (the 'not winning'). If water Qi is deficient then earth aids it (the 'not winning') and fire insults it (the 'winning').

This is further elaborated in the Nan Jing, wushisan nan shuo:—

'If a disease has empty evil, full evil, thief evil, minute evil and upright evil, how can they be distinguished? That which comes from behind is empty evil, that which comes from in front is full evil, that which comes from the not-winning is thief evil, that which comes from the winning is minute evil, and autogenous disease is upright evil.'

The quotation indicates the various paths taken by 'invading disease evils'. 'Empty evil' coming from 'behind' is a mother disease affecting the son as, for instance, a liver disease which is transmitted to the heart. 'Full evil' coming from 'in front' is a son disease going back to the mother, like a disease of the spleen which is transmitted to the heart. 'Thief evil' coming from the 'not-winning' can be illustrated by a liver disease being transmitted to the spleen and 'minute evil' coming from the 'winning' by a lung disease transmitted to the heart. 'Upright evil' is an autogenous heart disease, originating in the heart, and is not transmitted to any other organ (Fig. 36).

The following pages give some examples of the above laws, taken from the Zhongyixue Gailun with the Chinese phraseology, though it sounds a little strange to our ears, left for the most part intact.

Heart disease—palpitations and insomnia.

(a) If the fire of the heart is vigorous and abundant, the blood in the heart will be deficient, producing the following symptoms: restless sleep, an unquiet spirit, palpitations, constipation and ulcers of the tongue and mouth.

This type of insomnia originates in the heart itself; and, since it neither spreads to other organs nor results from their influence on

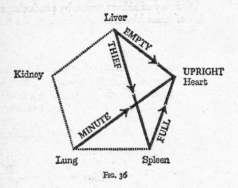

Fig. 36

the heart, it is treated directly by sedating the heart fire and tonifying the heart blood.

(b) If the spleen is empty and this emptiness affects the heart (son depriving mother of Qi), there will be a decrease of thirst and appetite, thin and watery stools, lethargy and weakness, palpitations, insomnia, amnesia etc.

In this case treatment of the heart would be ineffectual since the cause lies in the spleen. Therefore '*the spleen earth must be fortified, so that the spleen Qi becomes vigorous and does not deprive its mother of Qi; then the heart blood will be sufficient, the heart's spirit nourished and the illness cured.*'

(c) What in China are called 'empty exhaustion diseases' generally result from a deficiency of kidney Yin, causing 'empty fire' (the heart belongs to the element 'fire') to burn upwards. Fever ensues with copious and spontaneous sweating, coughing, vomiting of blood etc. and often insomnia. The insomnia is the result of a deficiency of kidney water; true Yin is not ascending and so the violence of the heart fire interferes with sleep.

Here the treatment is to strengthen the water and restrain the Yang, so that, when the kidney Yin becomes sufficient and the empty fire subdued, the disease will pass away.

These three examples show that heart disease, though the primary, is not the only cause of palpitations and insomnia. For a weakness of spleen earth or a deficiency of kidney water, by producing a corresponding deficiency of heart blood or a violence of the heart fire, may also occasion these symptoms.

Liver disease—headache and dizziness.

Among the commonest complaints caused by a dysfunction of the liver are headaches and dizziness. The condition is called 'liver fire ascending', because the patient often feels as if something hot were rising from the region of the liver into his head. In certain circumstances, however, these symptoms may be due to a deficiency of kidney water, a dysfunction of the lungs or inactivity of the spleen.

(a) If the liver Yang rises, wood (the liver belongs to the element 'wood') and fire will increase their activity, causing headache and dizziness, a flushed face, bloodshot eyes, a bowstring, unyielding pulse etc.

Since in this case the disease is in the liver and has not affected other organs, it is sufficient to treat the liver alone by sedating its excessive fire. Thereby the liver Yang will be balanced and the headache and dizziness will disappear.

(b) When headache and dizziness are due to a deficiency of water (kidney) and over-activity of wood (liver), the liver wind will rise and rush around in the head. The condition often occurs in those whose complexions are haggard and who experience the feeling of heat and emptiness. It is described by Shishi milu as one where *'kidney water is deficient and evil fire rushes into the brain'*.

According to the same source, *'if only the wind is treated, then the headache will increase and the dizziness become more pronounced. The appropriate treatment is full tonification of the kidney water, which will cause the symptoms to disappear.'*

In conformity with the teaching of the five elements, this phe-

nomenon is described by the saying, 'Water fails to submerge wood.'
The water of the kidney must be nourished so that it is able to submerge the wood of the liver; or, as the Chinese put it, 'If weak, then
tonify the mother.'

(c) Liver (wood) is under the control of lung (metal). (See Fig. 35.)
In a patient whose lung Qi (the word means either energy or breath)
is deficient, the air does not permeate the lungs so that the fluid in the
interstices, not being properly diffused and transformed, becomes
phlegm. In this condition the lungs are moist, the respiratory passages
become partially obstructed, there is coughing up of sputum and a
diminution of thirst and appetite. (In Chinese thought, food and
drink are associated with the stomach coupled to the spleen, and both
spleen and stomach with the element 'earth', which is endangered by
dampness and phlegm.) The accumulation of moisture and phlegm
causes also dizziness and a feeling of obstruction in the area of the
diaphragm. In this connection the Su Wen, xuanji yunbing shi,
says:-

*If liver wood is violent, it must be because metal is decayed and cannot
regulate it; and wood also creates fire.*

The best cure therefore for this type of dizziness is stimulating
earth to create metal, so that the lung Qi may circulate and the
liver wood become balanced.

These three examples show that, though headache and dizziness
are directly due to a disharmony of liver wood, they may also be the
results of a dysfunction of the lung, kidney, spleen and stomach.
The treatment is nourishing water to submerge wood, purifying the
liver to sedate fire and tonifying the lungs to regulate the liver.

Spleen diseases—diarrhoea.

Commonest among the many causes of diarrhoea is an empty
spleen, with its accompanying dampness. It can also result from a
deficiency of kidney Yang, where fire does not create earth, or from
a liver disease affecting the spleen.

(a) Weakness of the spleen Yang may produce uncontrolled
diarrhoea, with associated symptoms of a lack of appetite and thirst,
the desire to pass stool after eating, a feeling of obstruction and
melancholy in the lower chest and upper abdomen and a weakness
in the limbs. This type of diarrhoea may be cured by tonifying the
spleen.

Another type is caused by dampness and may be treated via the spleen in order to dry this out, in accordance with the advice of the Su Wen:- 'If dampness is excessive, then sedate.'

(b) One form of diarrhoea, in which the stool is variegated in colour, can occur because the 'fire of the gate of destiny' (see 'The Meridians of Acupuncture', ch. IX) is too weak and cannot create earth. It may be associated with symptoms of uneasiness in the stomach, anorexia and slight abdominal pain. The cause is described in the Yizong bidu thus:—

'*The kidneys, the roots of sealing and storing, control the two excretory passages; and, though they belong to the category 'water', true Yang lodges within them. A little fire creates Qi; and fire is the mother of earth. If this fire is decayed, how can the triple warmer, which belongs to ministerial fire, be activated or nourishment digested?*'

This is treated by tonifying fire to create earth, thereby restoring the kidney Yang; which, in actual practice, means tonifying the kidneys.

(c) If the liver wood is excessive, it may injure the spleen earth, causing abdominal pains, which are relieved by diarrhoea.

The cure will be incomplete if only the spleen is tonified or the liver sedated, for the abdominal pain is caused by the liver Qi rebelling, and the diarrhoea by the spleen Qi being empty.

Lung diseases—coughing and dyspnoea.

Coughing and dyspnoea are the two most characteristic symptoms of a lung disease but they are not in every instance directly attributable to this cause.

(a) If fluid from within the body collects in the lungs at the same time as chilly conditions prevail outside, there will be difficulty in breathing, fever, perhaps urinary trouble and a strong dislike for cold weather. In this instance the origin of the disease is in the lungs, which should be treated directly.

(b) Under the heading 'empty exhaustion disease' the Chinese include pulmonary tuberculosis, the general condition which often precedes this, and other conditions with similar symptoms. The patient will suffer from a chronic cough of long duration. The lungs will be empty and the spleen and stomach inactive, with consequent lack of appetite and watery stools.

The best treatment is to stimulate earth to create metal. Merely to tonify or moisten the lungs will often aggravate these symptoms, for the type of medicine that tonifies the lungs may obstruct the stomach and the type of medicine that moistens them may render the intestines slippery. One should first rectify the spleen and harmonise the stomach, thus restoring the balance in the function of these organs. There will then be an improvement in the appetite and the diarrhoea will cease. At the same time, if the lungs gain sustenance from the 'nourishment Qi' they will be naturally restored to health, and the coughing will be cured. The underlying principle of this treatment is that the spleen belongs to earth and the lungs to metal: so that earth is tonified to create metal.

(c) If the lungs (in the upper part of the body) are full and the kidneys (in the lower part of it) empty, the condition will cause coughing with much phlegm, aching of the loins, a fine pulse and, possibly, spermatorrhoea.

In this case both the lungs and the kidneys must be treated, for to treat only the fullness of the lungs might aggravate the emptiness of the kidneys, while to tonify only the kidneys might increase the fullness of the lungs. The lungs belong to metal, the kidneys to water and, because of their mutual interaction, should both be treated together.

(d) If the kidneys are empty and cannot transmit their Qi to the lungs, the result will be:— coughing, dyspnoea, a weak voice, shortness of breath, insufficient phlegm, breathlessness after exertion and possibly also an ache in the loins and increased urination.

To treat this condition one should tonify the kidneys so that they are able to transmit Qi. As we have seen, kidneys belong to water, lungs to metal. Normally, metal creates water, but in this case kidney water is deficient and thus unable to effect the basic transformation of lung metal. This disease therefore belongs to the category known as 'depriving mother of Qi' and the appropriate treatment is for the 'son to cause the mother to be full.'

(e) In a case where liver wood is violently active, and wood and fire blaze upwards, the inability of lung metal to descend will cause numbness of the throat with coughing and pain in the ribs.

In such a condition one should purify metal to' regulate wood. If liver wood is balanced, lung metal is not insulted by it and the disease will disappear. This treatment follows the law known as 'insulting backwards'.

Kidney diseases—spermatorrhoea

Spermatorrhoea, is due to a deficiency of kidney Qi. '*The kidneys receive the Jing (sexual power) from the five solid organs and the six hollow organs, and store it.*'

(Neijing)

Thus the principal treatment of spermatorrhoea is so to tonify the kidneys that they can store and control Jing. There are, however, other causes of the disease.

'*Each of the five solid organs has its separate function. As long as Jing is stored, there is health; if one of the solid organs is in severe disorder, it will weaken the control of the heart and kidneys over the Jing*'.

(Yixue rumen)

This quotation shows that, apart from an empty kidney causing the Jing gate to be weak, abnormality in some other organ may have the same effect, such as excessive heart fire, dampness and heat in the liver meridian, emptiness of the heart as well as kidneys or failure of the reciprocity of water and fire.

(a) Certain people are weak by nature from birth, sometimes because they were born of elderly parents, and will therefore suffer from a deficiency of true Qi in the kidneys. They will not infrequently have nocturnal emissions, with aching of the loins, dizziness and noises in the ears. This condition, being due to a weakness of the kidneys, is therefore treated by direct stimulation of them. .

(b) Spermatorrhoea may occur in patients who suffer from constant anxiety, frustration, worry and dreaming.

The appropriate treatment here is to purify fire in order to pacify water. If heart fire is balanced, kidney water will be restored to tranquillity and the patient cured; but, if mistakenly the kidneys are tonified in order to strengthen the Jing gate, the treatment will not only be ineffective but may even aggravate the condition.

(c) Sadness and depression, associated with over-abundance of liver fire, may cause spermatorrhoea.

This may be treated by purifying and sedating the liver fire. Because the kidneys control 'closing and storing', and the liver manages 'clearing and purging', violence of liver fire will result in excessive clearing and purging, which can affect the closing and storing of the kidneys and so cause spermatorrhoea. Although

88   ACUPUNCTURE

purifying and sedating the liver fire will not directly cure spermatorrhoea, balancing the liver fire to reduce the clearing and purging will restore the kidneys naturally to their proper function. This is what is meant by: 'If full, sedate the son.'

(d) A general physical weakness will often produce aching of

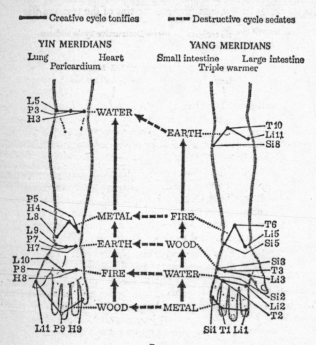

━━━━ Creative cycle tonifies   ▬ ▬ ▬ Destructive cycle sedates

YIN MERIDIANS   YANG MERIDIANS

Lung   Heart   Small intestine   Large intestine
     Pericardium   Triple warmer

Fig. 37

the loins, lack of strength in the legs, dreaming, insomnia and nervousness. This is usually caused by deficiency of the heart and kidneys. Patients will also often suffer from spontaneous sweating and spermatorrhoea because, kidney water being absent and heart fire disquieted, water and fire are not co-operating.

The appropriate treatment is to cause water and fire to assist one another, so as to restore the relation between kidneys and heart.

The Five Element Acupuncture Points

The general classification of the five elements is worked out in greater detail for those acupuncture points that lie between the fingertips and the elbow, and between the tips of the toes and the

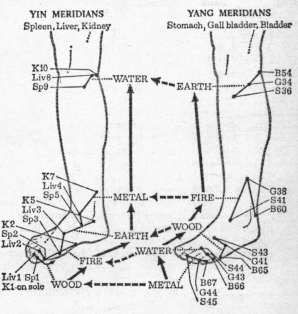

FIG. 38

knees, a series of points whereby it is possible to treat nearly any disease, wherever it may appear on the body, without having recourse to any other acupuncture points (Fig. 37 and 38).

It will be noted that the direction of movement is centripetal, from the finger or toe tips inwards in accordance with the creative cycle, i.e., from the finger tip, anterior surface, wood, fire, earth, etc. Equivalent positions on the anterior and posterior surfaces are by the law of five elements, antagonistic to one another, i.e., at the finger tip metal and wood are antagonistic in such a manner that the external surface (metal) is the destructive agent. Thus both the creative and the destructive cycles are centripetal, moving from without inwards, i.e. from the tips of fingers or toes to the elbow or knee, (creative cycle); or from the posterior surface, which is embryologically an external surface, to the anterior surface, which is embryologically a more internal surface, (destructive cycle).

It will also be noted that the arrangements of the elements on the anterior surface of the arm is the same as that on the medial (embryologically anterior) surface of the leg, and, similarly, the arrangement of the elements on the posterior surface of the arm are the same as those on the lateral (embryologically posterior) surface of the leg.

This arrangement of the five elements on the limbs is fundamental to the arrangement of the points of tonification and sedation:

The point of tonification of the meridian and element to be tonified, is a 'mother' (i.e. preceding) point.

The point of sedation of the meridian and element to be sedated, is a 'son'. (i.e. following) point.

e.g. 1. The heart meridian is a fire meridian. The fire point of the heart meridian is H8. According to the five element theory, wood is the mother of fire, so that, if the wood element in the fire meridian is stimulated, the 'son' fire, would be tonified. Therefore, the point of tonification of the heart meridian is H9 (the wood point on a fire meridian).

2. The 'son' of the element fire is the earth. If the 'son' is stimulated he takes energy from his 'mother' who, as a result, is weakened. Therefore the point of sedation of a fire meridian is its earth point which, in the case of the heart meridian, is H7 (earth point on a fire meridian).

3. The gall bladder meridian is a wood meridian. The 'mother' of wood is water, so that the point of tonification would be the water point G43. The 'son' of wood is fire, so that the point of sedation would be the fire point on this Yang wood meridian, point G38.

YIN meridians:

Organ	Element	Wood	Fire	Earth	Metal	Water
Lung	Metal	L11	L10	L9	L8	L5
Heart	Fire	H9	H8	H7	H4	H3
Pericardium	Fire	P9	P8	P7	P5	P3
Liver	Wood	Liv1	Liv2	Liv3	Liv4	Liv8
Spleen	Earth	Sp1	Sp2	Sp3	Sp5	Sp9
Kidney	Water	K1	K2	K5	K7	K10

YANG meridians:		Metal	Water	Wood	Fire	Earth
Large intestine	Metal	Li1	Li2	Li3	Li5	Li11
Small intestine	Fire	Si1	Si2	Si3	Si5	Si8
Triple warmer	Fire	T1	T2	T3	T6	T10
Gall bladder	Wood	G44	G43	G41	G38	G34
Stomach	Earth	S45	S44	S43	S41	S36
Bladder	Water	B67	B66	B65	B60	B54

Treatment via Law of Five Elements

The creative and the destructive cycles of the law of the five elements may be used together or separately in the treatment of disease:

1. If the pulse of the stomach shows an overactivity, as is usually the case with a hypersecretion of acid, with resultant duodenal or stomach ulcers, the following will usually correct it:

(*a*) The stomach meridian belongs to the element earth which should be sedated after the 'mother-son' law. Metal is the 'son' of earth. The metal point of the earth Yang meridian (stomach) is S45—which is sedated.

Also the metal point of the metal Yang meridian (Large intestine) is sedated (in order to further drain the energy from earth) which is Li1.

(*b*) The gall bladder is opposed to the stomach by the law of the five elements. This is therefore tonified at its wood point (to destroy earth), which is G41.

Similarly the wood point of the stomach meridian itself is tonified as this destroys the element earth within the stomach meridian itself. This is S43.

2. Likewise, if the pulse of the stomach is underactive:

(*a*) The 'mother' must be tonified. The 'mother' of the Yang earth (stomach) is Yang fire (small intestine). Hence the fire point in the stomach meridian itself is tonified, point S41.

Also the fire point in the fire Yang meridian (small intestine) itself is tonified, point Si5.

(b) The opposed element, wood, must be weakened, so that the element (earth, stomach) that it opposes is, as a result, strengthened. Hence the wood point of the stomach meridian itself is sedated, point S43.

Also the wood point of the wood (Yang) meridian (Gall bladder) itself is sedated, point G41.

It will be noted that in this context the law of the five elements only operates within elements belonging to either Yin or Yang, i.e., if an organ is a Yin organ its 'mother', 'son' and opposed element are also Yin. Sometimes this is not the case: a Yin organ may effect a Yang organ, and a Yang organ a Yin organ.

Continuing this line of reasoning, a full table of what should be done in each circumstance is therefore as follows:

Organ	To Tonify, i.e. if organ is underactive				To Sedate i.e., if organ is overactive			
	Tonify		Sedate		Sedate		Tonify	
Lungs	L9	Sp3	L10	H8	L5	K10	L10	H8
Kidney	K7	L8	K5	Sp3	K1	Liv1	K5	Sp3
Liver	Liv8	K10	Liv4	L8	Liv2	H8	Liv4	L8
Heart	H9	Liv1	H3	K10	H7	Sp3	H3	K10
Spleen	Sp2	H8	Sp1	Liv1	Sp5	L8	Sp1	Liv1
Large intestine	Li11	S36	Li5	Si5	Li2	B66	Li5	Si5
Bladder	B67	Li1	B54	S36	B65	G41	B54	S36
Gall bladder	G43	B66	G44	Li1	G38	Si5	G44	Li1
Small intestine	Si3	G41	Si2	B66	Si8	S36	Si2	B66
Stomach	S41	Si5	S43	G41	S45	Li1	S43	G41
Pericardium	P9	Liv1	P3	K10	P7	Sp3	P3	K10
Triple warmer	T3	G41	T2	B66	T10	S36	T2	B66

It will be noticed that amongst the four points that should be used in tonifying an organ, is the point of tonification; likewise amongst the four points that should be used in sedating an organ is the point of sedation.

For this reason the law of the five elements is quite often used if the simple points of tonification or sedation do not work, or only do so partially.

Likewise the law of the five elements may be used if it is noticed that the initial illness has various complications corresponding to the other factors mentioned under each element. e.g., If a patient has predominately a cardiac disease which is due to an under-activity of the heart and has as complications a malfunction of the liver and kidney, the choice of this law is obvious, for underactivity of the heart not only is the heart treated, but also the liver and kidney—as may be seen from the chart, points H9, H3 (heart), Liv1 (Liver), K10 (Kidney).

Extension of the Law of Five Elements

We have already seen that the basis of Chinese traditional medicine is the polarity of Yin and Yang, the negative and the positive, in addition to which is the division of the body in accordance with the twelve primary organs and meridians. But, forming a bridge between these two groups, is a third: that of the five elements.

Six of the twelve organs and meridians are Yin, six are Yang, and each of the five elements controls one Yin organ and meridian and one Yang, except the element 'fire', which in both groups controls two. In the Yin group, the heart belongs to 'princely fire', the pericardium to 'ministerial fire'; in the Yang group, the small intestine belongs to 'princely fire,' the triple warmer to 'ministerial fire'.

The six Yin organs are referred to in Chinese literature as the five Zang (or solid) organs; the six Yang organs are called the six Fu (or hollow) organs and are mentioned in the following quotation:—

'What are called the five solid organs store life essence and energy (Jing and Qi) and do not let them leak away; therefore they are filled but cannot be full. The six hollow organs transmit and transform matter but do not store it; thus they are full but cannot be filled.'

(Su Wen, wuzang bielun)

The Chinese consider that each organ has an effect on and interacts with a bodily tissue, sense organ, season etc., as described below.

Element	Wood	Fire	Earth	Metal	Water
Yin organ	Liver	Heart	Spleen	Lungs	Kidney
Yang organ	Gall bladder	Small intestine	Stomach	Large intestine	Bladder
Sense commanded	Sight	Words	Taste	Smell	Hearing
Nourishes the	Muscles	Blood vessels	Fat	Skin	Bones
Expands into the	Nails	Colour	Lips	Body hair	Hair on head
Liquid emitted	Tears	Sweat	Saliva	Mucus	Urine
Bodily smell	Rancid	Scorched	Fragrant	Fleshy	Putrid
Associated temperament	Depressed	Emotions up & down	Obsession	Anguish	Fear
	Anger	Joy	Sympathy	Grief	
Flavour	*Sour	Bitter	Sweet	Hot	Salt
Sound	Shout	Laugh	Sing	Weep	Groan
Dangerous type of weather	Wind	Heat	Humidity	Dryness	Cold
Season	Spring	Summer	Mid-summer	Autumn	Winter
Colour	Green	Red	Yellow	White	Black
Direction	East	South	Centre	West	North
Development	Birth	Growth	Transformation	Harvest	Store
Beneficial cereal	Wheat	Millet	Rye	Rice	Beans
Beneficial meat	Chicken	Mutton	Beef	Horse	Pork
Musical note	chio	chih	kung	shang	yu

I myself have not been able to verify all the factors mentioned in the above chart, and I have my doubts about the correctness of a few of them, but that the majority are correct I have no doubt as I use them in diagnosis and treatment with success.

It is known for example that someone who has a liver (wood) weakness is more sensitive than the average person to an East (wood) wind (wood), that his nails (wood) may be blemished and that he may have foggy vision with black spots (wood). That the person who feels cold (water) in his bones (water) is found by the

*Sour like vinegar, bitter like bitter lemon, sweet like sugar, hot like ginger, salt like common salt.

pulse diagnosis to have a weakness of the kidneys (water). That the person with verbal (fire) diarrhoea and a high colour (fire) has an overactivity of the heart (fire). That the frightened (water) child who does not want to sleep in the dark and wets (water) his bed has a weakness of the kidney (water). That the diabetic (earth) who eats too much sugar (earth) will probably end up with a diabetic (earth) coma, or the renal (water) hypertensive who eats too much salt (water) may finish in the grave.

The Zhongyixue Gailun gives the following example of the interdependence of various factors within one element (in this case, wood). In spring, all plants, as they grow and put forth their fresh green shoots, are visible manifestations of the luxuriant Qi of birth. Wood, therefore, represents the season of spring. Of the five developing processes it belongs to birth; of the five climatic conditions, to wind. In the physical body, both wood and spring are represented by the liver, whose nature is happy, straightforward and cheerful. Closely connected to the gall bladder, the liver influences the eyes and controls the muscles, which is why liver diseases are usually revealed by symptoms of disturbed vision and muscular spasm. People with too violent a liver are liable to fits of rage and, conversely, fits of rage are liable to cause injury to the liver. Among the five emotions, therefore, anger is under the control of the liver. The colour green is associated with wood, and it is worth noting that many liver diseases are recognisable by a greenish tinge in the patient's skin.

Indirect effects, which are more rarely found must not be forgotten: Foggy vision (wood) is normally due to an underactivity of the liver (wood), but occasionally it may be caused by an underactive kidney (water) [which is the 'mother' by the law of five elements.]

This system may also be used therapeutically:

In psychology a person who has an endogenous depression (wood) may be cured by treating the liver (wood). Or someone who weeps (metal) a lot may after the destructive cycle of the five elements be told to laugh (fire) more, which quite naturally would stop the weepiness. If just being told to laugh (fire) is not enough, the heart (fire) itself may be stimulated either by acupuncture or by giving the correct cardiac (fire) tonic or by eating bitter (fire) food—though as a rule the more powerful acupuncture works best.

Case History. A patient was seen who was unable to stop talking (words-fire). The pulse diagnosis revealed an overactivity of the pulse of the heart (fire). The 'son' of the fire is the earth. Therefore to sedate the fire the earth point (H7) of the fire (heart) meridian was used. Within a few minutes of the needle being in place the verbal diarrhoea stopped and the patient spoke normally for about a day when the incessant flow of words started again. Similar treatment was repeated to effect a cure.

Case History. A patient had nails (wood) which kept on cracking and were thin and brittle with longitudinal ridges (calcium had been tried to no avail). Her eyes (wood) watered very easily especially in the wind (wood). Her body had a slightly rancid (wood) smell and she easily became angry (wood) when she would shout (wood) a lot. The pulse diagnosis revealed an underactivity of the liver (wood). The 'mother' of wood is water. Therefore the water point (Liv8) on the liver meridian was used. This or similar treatment was repeated to effect a cure.

It is interesting to note that the improvement in the nails started within two weeks of initiating treatment. Yet a nail takes some four months to grow up from its base, suggesting that a little revision is needed in the theory of the physiology of nails. This observation has been made repeatedly with various patients: the time of response may vary, but is always faster than would be expected from the speed of growth of the nail.

Passages from certain Chinese texts discuss in detail the correlation of the five elements shown in the chart a few pages back.

'*The liver creates the muscles, the muscles create the heart… the heart creates the blood, the blood creates the spleen… the spleen creates the flesh, the flesh creates the lungs… the lungs create the skin and body air, the skin and body air create the kidneys… the kidneys create the bone marrow, the bone marrow creates the liver.*'

(Su Wen, yingyang yingxiang dalun)

The reciprocal relationship among the five (or six) solid organs is as follows:—

'*The kidneys are the controller of the heart, the heart is the controller of the lungs, the lungs are the controller of the liver, the liver is the controller of the spleen, the spleen is the controller of the kidneys.*'

(Su Wen, wuzang shengcheng lun)

Certain organs, mainly hollow, have the function of transmitting and transforming:—

'*Spleen and stomach, large intestine, small intestine, triple warmer and bladder are the root of the granaries, the dwelling-place of nourishing Qi, and they are called organs. They are able to transform the dregs and transmit tastes so that they can enter and depart.*'

(Su Wen, liujie zangxiang lun)

The alternating function of the organs, so important a principle in Chinese medicine, is thus referred to:—

'*If the stomach is filled, then the intestines are empty; if the intestines are filled, then the stomach is empty; they are alternately filled and emptied, so that energy is able to ascend and descend.*'

(Ling Shu, pingren juegu pian)

Coupled organs, or Yin and Yang organs belonging to the same element are also mentioned:—

'*The bladder and kidney are outside and inside, the gall bladder and liver are outside and inside, the stomach and spleen are outside and inside. These are the Yang and Yin organs and meridians of the leg. The small intestine and heart are outside and inside, the triple warmer and pericardium are outside and inside, the large intestine and lung are outside and inside. These are the Yang and Yin organs and meridians of the arm*'

(Su Wen, xueqi xingzhi pian)

There is also a relationship between the solid organs and the limbs:—

'*If the lungs or heart have an evil, its Qi remains in the two elbows; if the liver has an evil, its Qi remains in the two armpits; if the spleen has an evil, its Qi remains in the two thighs; if the kidneys have an evil, its Qi remains in the knees.*'

(Ling Shu, xiekepian)

Apart from the relationship between what the Chinese called 'the eight hollows' (elbow, armpit, groin, knee) and the solid organs, the spleen has a special function in relation to the limbs:—

'*The Emperor asked, "Why is it that with a disease of the spleen the four limbs are useless?" Qi Bo, the Emperor's physician, replied, 'The four limbs all take Qi from the stomach; but, if it cannot get through, then it must be taken from the spleen.*"'

(Su Wen, taiyin yangming lun)

Further interconnections, which may be useful in diagnosis and for an understanding of Chinese physiology, are illustrated below:—

'That which combines with the heart is the blood vessels, and its flourishing is the colour of the complexion... That which combines with the lungs is skin, and its flourishing is the body hair... That which combines with liver is muscle, and its flourishing is the nails... That which combines with the spleen is flesh, and its flourishing is the lips... That which combines with the kidneys is the bones, and its flourishing is the hair on the head.'

(Su Wen, wuzang shengcheng lun)

The relationship is extended to the orifices of the body:—

'The liver is related to the holes of the eyes, the heart is related to the holes of the tongue, the spleen is related to the hole of the mouth, the lung is related to the holes of the nose, the kidneys are related to the holes of the ears.'

(Su Wen, yinyang yingxiang dalun)

The five emotions are also correlated:—

'Man has five solid organs, which transform the five Qi to create anger, joy, over-concentration, anguish and fear.'

(Su Wen, yinyang yingxiang dalun)

In detail: the liver is injured by anger, the heart by joy, the spleen by over-concentration, the lungs by anguish and the kidneys by fear.

It is a well-known fact that the colour of the face reflects the inner condition of the body; the alcoholic can be recognised by his florid complexion, the anaemic by his pallor; the skin of the hepatic is dusky or pasty, that of the jaundiced a distinctive yellow. The Chinese classify the colouring of the whole face, including eyes and eyebrows, in a traditional pattern: green indicates a hepatic disease, red a cardiac disease, yellow a splenic disease, white a pulmonary disease and black a renal disease. The colours may, however, be changed by some indirect effect. The patient with a liver disease may be white rather than green, because metal destroys wood; the cardiac patient black rather than red, because water destroys fire.

Clearly this colour system cannot be taken too literally. It should be regarded rather as a guide to very slight differences in colour, so slight that a doctor, aware of them at a first quick glance, might fail to perceive them at all, if he stared too hard at his patient. They are like the subtle variations of colour revealed to the artist when he sees a delicate harmony of reds, yellows and blues in a cloud, which the average man would describe simply as 'white or 'grey'.

The five tastes (sour, bitter, sweet, hot and salty) are related to their corresponding organs:—

'Each of the five tastes moves to what it likes. If the taste of the nourishment is sour, it moves first to the liver; if bitter, it moves first to the heart; if sweet, it moves first to the spleen; if hot it moves first to the lungs; if salty, it moves first to the kidneys.'

(Ling Shu, wuwei pian)

According to this theory, if food of the same type of flavour is continuously eaten, the corresponding organ will be overloaded and disease will result:—

'Sour injures muscles, bitter injures energy (Qi), sweet injures flesh, hot injures skin and body hair, salt injures blood.'

(Su Wen, yinyang yingxiang dalun)

Health is best maintained by a diet of balanced flavours.

Theoretically it should be possible to tonify or sedate a malfunctioning organ by a one-sided diet, increasing or decreasing one or other of the five flavours so as to stimulate the organ. While I was in China, I enquired of many doctors whether this was in fact their practice, but I never heard of anyone who had tried it. The law seems to be applied only in the positive sense of preserving health by a mixture of the flavours. This theory should be compared with that in chapter X at the end of the section 'excess of food and drink.'

The five elements and organs are also associated with meteorological conditions:—

'The heart communicates with the summer Qi, the lungs with the autumn Qi, the kidneys with the winter Qi, the liver with the spring Qi, the spleen with the earth Qi (i.e. late summer)'

But the meteorological conditions could also be the cause of disease:—

'In spring, if one is injured by wind, the evil Qi remains and causes diarrhoea; in summer, if one is injured by the heat, it causes intermittent fever during the autumn; in autumn, if one is injured by the damp, it rebels upwards and causes coughing and paralysis; in winter, if one is

injured by the cold, warm diseases are inevitable in the spring. Thus the Qi of the four seasons injure the five solid organs.'

(Su Wen, shengqi tongtian lun)

Some inconsistencies can be observed in this quotation.

The evolution of the whole of life was governed not only by the division into the Yin and Yang of birth and death but also by the division into the five elements of creation, growth, change, gathering and storing. In the botanical sense, this would mean that the season of creation, when plants put forth their first shoots, is spring ('wood'); their main period of growth is during the summer ('fire');' they begin to change their colouring in late summer ('earth'); their fruit is gathered in autumn ('metal') and their seeds or bulbs are stored beneath the earth in winter ('water'). The evolution of man and animal traces the same pattern: infancy ('wood'), youth ('fire'), adulthood ('earth'), decline ('metal') and death ('water').

'Thus the Yin and Yang seasons are the beginning and end of the myriad things, the root of birth and death. If they are disobeyed, calamity will result; if they are obeyed, disease will not arise.'

(Su Wen, siqi tiaoshen dalun)

VII

LAWS OF ACUPUNCTURE

The organ-meridians that are made use of in acupuncture are not isolated, nor do they function entirely independently of one another.

It is well known in standard medical practice that if, for example, the heart is tonified, whether it be by medicinal or other means, there are other secondary indirect effects which ensue. These secondary effects, whether they be wished for or not, are a more or less unavoidable sequence of the primary stimulus.

The main law discussing the interrelationship of these primary and secondary effects, is that of the five elements, largely mentioned in the previous chapter. This chapter will deal with other inter-relationships and some applications of the law of five elements.

It is inadvisable to treat patients by acupuncture unless possible secondary effects are clearly visualised and, if necessary, avoided or corrected by secondary treatment, for occasionally in a hyperacute state or in an over-sensitive patient, the reaction obtained may be too acute. In this case the diseased organ is treated indirectly, e.g. for a disease of the heart the gall bladder is treated.

Chinese tradition expressed the interrelations between the various meridians and the methods of treatment designed to take advantage of, and to avoid undesired ill-effects which might result from, the intimate connections between organs, in a series of laws. These, formulated in the symbol imagery characteristic of Chinese thought, are as follows:

The 'Mother-son' law.

The 'Husband-wife' law.
The 'Midday-midnight' law.
.etc.

The 'Mother-son' law

The essence of this law is set out in the Chinese text as follows:

'If a meridian is empty, tonify its mother. If it is full, disperse the child.'
(Zhenjiu Yixuè)

As the Qi (in the superficial circulation of energy) flows through the meridians in a certain order, the preceding organ (the 'mother') receives the energy first and gives it on to that which follows (the 'son'). In the case of excess or deficiency of one to two such related organs, it is frequently preferable to give treatment via the 'mother' of the affected meridian, rather than directly.

This law has various applications:

A. SUPERFICIAL CIRCULATION OF ENERGY

As mentioned Qi flows through the meridians in a certain established order, i.e. from lung to large intestine, stomach, spleen, heart, etc.

Here the lung functions as the 'mother' of the large intestine; i.e. the large intestine is the 'son'. Or again, the large intestine is the 'mother' of the stomach; i.e. the stomach is, in this case, the 'son'. In this example the large intestine may function either as the 'son' or the 'mother', depending on whether it is related to the preceding or the following meridian respectively.

The flow of Qi to the 'son' is dependent on that of the 'mother'. Therefore, if the large intestine is tonified, so that the large intestine (in this case the 'mother') has more energy, its child, the stomach, will, as the energy flows on, also receive more energy i.e. be tonified. As the 'mother' (large intestine) is full of energy the flow of Qi coming from the preceding meridian (the lung) is dammed back, so that there is also an increase of energy in the lungs. Hence:

Tonification of the 'mother'—(Large intestine)

produces tonification of the 'son'—(Stomach)

and secondarily tonification of the preceding meridian—(Lung)

The effect on the preceding meridian (in this case the lung) is usually less marked than the true 'mother'—'son' effect.

If correspondingly the 'mother' meridian is sedated instead of tonified the result will be:

'Mother' meridian	— sedated
'Son' meridian	— sedated
Preceding meridian	— sedated

The process involved is the same in reverse.

B. DEEP CIRCULATION OF ENERGY

According to the plan of the pulse diagnosis the organs follow one another in a certain order. This order is the same as that used for the pulse diagnosis at the wrist, which will be described in a later chapter.

THE ORDER OF PULSES

Left radial artery			*Right radial artery*
Superficial	*Deep*	*Deep*	*Superficial*
Small intestine	Heart	Lung	Large intestine
	↑		↑
Gall bladder	Liver	Spleen	Stomach
	↑		↑
Bladder	Kidney	Pericardium	Triple warmer

In detail this would be:

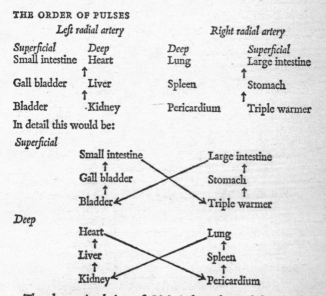

Superficial

Small intestine · · · · · · · · · · · Large intestine

Gall bladder · · · · · · · · · · · Stomach

Bladder · · · · · · · · · · · Triple warmer

Deep

Heart · · · · · · · · · · · Lung

Liver · · · · · · · · · · · Spleen

Kidney · · · · · · · · · · · Pericardium

The above circulation of Qi is independent of the circulation of Qi in the better known superficial circulation of energy. The energetics of deep circulation of energy is often noted in the pulse diagnosis when a weakness of the kidney, if it has persisted long enough, is accompanied by a weakness of the liver.

Some examples of the deep circulation of energy and the 'mother' —'son' law:

1. Tonification of the kidney ('mother') produces,
   Tonification of the liver ('son') and,
   Tonification of the lungs (preceding organ).
2. Sedation of the large intestine ('mother') produces,
   Sedation of the bladder ('son') and,
   Sedation of the stomach (preceding organ).

It will be noted that when various laws operate at the same time, certain effects are additive while others cancel each other out, or, on the other hand, the more powerful effect may dominate.

The deep circulation of energy is exactly the same as the creative cycle of the five elements for both the Yang and Yin organs— perhaps an unnecessary complication of saying the same thing in two ways. The only additional information given, is that it shows that the triple warmer is the son of the small intestine, and the pericardium the son of the heart.

The 'Husband-wife' Law

Organs which have equivalent positions on the left or right pulse are related to one another by the law, called in Chinese, 'husband'— 'wife'. They are:

Left wrist—'Husband'— Yin		Right wrist—'Wife'— Yang
Small intestine		Large intestine
Heart		Lung
Gall bladder	Dominates	Stomach
Liver		Spleen
Bladder	Puts in danger	Triple warmer
Kidney		Pericardium

There is thus a relationship, for example, between the heart and lungs.

The pulses on the left wrist are considered Yin while those on the right are Yang. This fits in the general conception that the right hand is the active hand—the hand that holds the bow of the violin, that throws the ball, etc., so it is not surprising that it is the right hand which functions as the Yang.

The pulses on the left are considered the 'husband' while those on the right are the 'wife'. This is the opposite of what one might expect as the 'husband' is the male principle (Yang) and is therefore on the right-hand side of the body, the reverse being true for the 'wife'. This is a conception that is often met: namely, that polar opposites are at the same time opposed to one another in function, and yet constitute a single unit. e.g. In the upper half of the body the right hand is usually the more powerful and is the doing hand, while in the lower half of the body it is the left foot which is more active, setting the rhythm in marching or in beating time to the music. The marching soldier keeps time in the rhythm between his left foot and right hand.

The 'husband' (left wrist), is said to dominate the 'wife', while the 'wife' (right wrist) is said to contribute stability and solidarity.

The pulses on the left ('husband') should be slightly stronger than those on the right ('wife').

'Weak "husband", strong "wife"; then there is destruction. Strong "husband", weak "wife"; then there is security.'

(Zhenjiu Dacheng)

The 'Midday-midnight' law

Qi takes its course through the twelve meridians, as mentioned before, over a period of twenty-four hours.

According to the law based on this daily rhythm there is a relationship between organs which receive their maximal flow at opposed times. e.g. The heart, which has its maximal activity at twelve midday, and the gall bladder, which has its maximal activity at twelve midnight (Fig. 39). The relationships of the heart and gall bladder are well known to Western medicine:

1. A patient presenting symptoms of typical angina pectoris with the usual electrocardiographic findings, may in reality have

biliary colic, the cardiac symptoms, although they are more severe, being secondary to the law 'midday—midnight', to the gall bladder.

2. I have noticed that if I have treated a patient's heart by acupuncture rather too powerfully, that he may have biliary colic, lasting for about half an hour round about midnight of the same day.

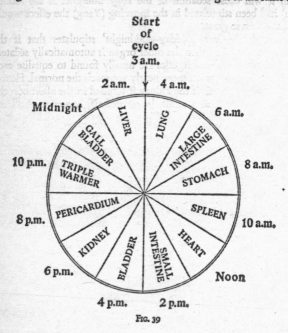

FIG. 39

3. Possibly the relation (if it is true) between angina pectoris and the continual excessive eating of saturated fatty acids or carbohydrates may one day be explained by means of a lipolytic function of bile or the metabolism of the liver.

The 'midday—midnight' law is applied as follows:

If an organ is stimulated by a moderate stimulus only the organ itself is effected. If the same organ is strongly stimulated, the organ with which it is connected by the law 'midday—midnight', is stimulated in the opposite sense. This law is more effective if a Yin organ is stimulated at a Yin time (midday to midnight) and if a Yang organ is stimulated at a Yang time (midnight to midday).

e.g. If the kidney is tonified in the afternoon (Yin organ, Yin time) it will cause sedation of the large intestine. If the kidney (Yin) had been stimulated in the morning (Yang) the effect would not have been so great.

Although the law 'midday—midnight' stipulates that if the one organ is tonified then the other organ is automatically sedated, in actual practice the energetics are usually found to equalise each other so that both organs more nearly approach the normal. Hence, in the above example, if the kidney is tonified in the afternoon the large intestine will, as a result, be sedated. If, however, the large intestine were already in an under active state, it would be tonified by tonifying the kidney.

Physiological Relationships

There are relationships between the various organs which are not covered by the actual laws of acupuncture, but a physiological relationship between those in question is obvious at a glance. They are mentioned in Zhenjiu Dacheng (II, p. 18v):

Liver	to help its function, sedate the large intestine
Large intestine	If ill, tonify the liver
Spleen	If ill, disperse the small intestine
Small intestine	If ill, disperse the spleen

Relation of Meridian and Region of Body

A relationship between the meridians treated and a region of the body is part of ancient tradition. It is rarely used in practice today.

Ling Tchou, Tsa tcheng loun stated in 250 B.C.:

'For diseases of the upper part of the body stimulate, above all, the meridian of the large intestine.

'For diseases of the central part of the body the meridian of the spleen.

'For diseases of the lower part of the body, the meridian of the liver.

'For diseases on the front of the chest, the meridian of the stomach.

'For the back, the meridian of the bladder.

'This is the most important part of the secret doctrine.'

THE MAIN CATEGORIES OF ACUPUNCTURE

The first section, chapter VI, section 'acu...
element, following this resolution, gives the po...
all acupoints, which is printed below. The po...
on the meridian whose function it controls it f...

VIII

THE MAIN CATEGORIES OF ACUPUNTURE POINTS

The thousand or so acupuncture points may be divided into various categories, all points in each category having similar properties.

Points of Tonification

The point of tonification of a meridian is the 'mother' point of its own element, i.e. the liver belongs to the element wood. The 'mother' (preceding element) of wood is water. Therefore the point of tonification is the water point—liver 8 (see chapter VI).

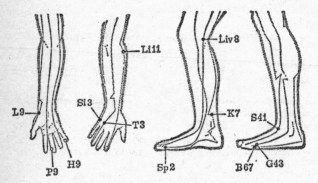

FIG. 40 Points of tonification .

The first column in chapter VI, section 'treatment via law of five elements,' following this reasoning, gives the points of tonification of all meridians, which is printed below. The point of tonification is on the meridian whose function it controls (Fig. 40).

Meridian	Point
Lungs	point L9 (which is also the source)
Large intestine	point Li11
Stomach	point S41
Spleen	point Sp2
Heart	point H9
Small intestine	point Si3
Bladder	point B67
Kidney	point K7
Pericardium	point P9
Triple warmer	point T3
Gall bladder	point G43
Liver	point Liv8

When a point of tonification is stimulated various results take place, some direct and some indirect. The indirect effects are as a rule of secondary importance, so that sometimes they may be neglected; though it can happen that they are of such importance that they even overshadow the primary effect of the direct tonification of the meridian, there being no appreciable direct effect and an overwhelming indirect effect. It is for this reason that all the direct and indirect effects must be taken into account with each and every point stimulated, so that no unwished for results occur. The conditions revealed by the pulse diagnosis will decide which of the various actions and reactions will take place.

DIRECT RESULTS

(a) The meridian itself is tonified, e.g. if the point Heart 9 (H9) is tonified, the meridian of the heart is tonified. This is shown by a greater strength of the pulse of the heart.

This effect is by far the most important of all the effects produced by stimulating a point of tonification (e.g. H9).

INDIRECT RESULTS

(b) The meridian in rapport with the meridian tonified by the law 'husband-wife' is sedated if it is in excess, e.g. if the heart is tonified, the lung is sedated. This is because as a rule not much energy is created de novo; the deficiency in the heart is made up by passing a part of the excess of energy in the lung over to the deficiency in the heart; so that whereas before treatment the heart was deficient and the lung in excess, after the stimulation of point H9, both the lung and the heart meridians have an equal amount of energy.

This law only operates in cases where the 'wife' (the 'husband' H9, is stimulated) has an excess of energy. If the 'wife' (the lung in this case) has the same amount of energy as the 'husband' (heart), or even less energy than the heart, this indirect effect does not take place.

(c) The meridian in rapport with the meridian tonified by the law 'midday-midnight' is sedated, if it is in excess and if treatment is given at the correct time of the day.

The heart has its maximal energy at midday, the gall bladder has its maximal energy at midnight. All organs are connected with their opposite number, from which they are separated by twelve hours, by a secondary meridian through the medium of which the mechanism of the law 'midday-midnight' takes place. If the heart is tonified the gall bladder is therefore sedated.

This law only operates to any marked degree if:

1. The opposite element (gall bladder) has an excess of energy.

2. The meridian tonified (heart) is tonified at a period of the day which is in its own sign. In this case, the heart which is a 'solid' organ and therefore Yin, must be tonified at a Yin period of the day which is from midday to sunset.

Thus if the heart is tonified at H9 in the afternoon (time Yin), the gall bladder will, as a result, be sedated. This would not have happened at all and, if at all, to a lesser degree if H9 had been tonified in the morning or if the gall bladder had had the same amount or less energy than the heart.

Conversely, if the gall bladder were tonified at point G43 (its point of tonification) in the morning, the heart would be sedated (provided the heart were also in excess).

(d) Between certain meridians there are special connections via secondary meridians, that do not follow strict laws, and are not invariably operative.

In this case if the heart is tonified at H9, the vessel of conception is sedated.

(e) Relationship between meridians sometimes operate via the superficial circulation of energy.

If a meridian is tonified, the meridian which comes before it and after it in the order of the superficial circulation of energy is tonified. That is to say, if the heart is tonified at H9, the spleen and the small intestine meridians will be tonified. (According to Niboyet the tonification of the meridian before and after the one stimulated is preceded by a rapid sedation, as the energy from these two encircling meridians first rushes to the tonified meridian, before being tonified themselves by the excess).

(f) Tonification of a meridian entails tonification of the meridian before it and after it in the order of the circulation of the deep flow of energy in accordance with the plan of the pulse. In this case tonification of the heart would also cause tonification of the liver and pericardium meridians. But this effect is not as marked as that of the superficial circulation of energy. (Similarly to the previous example (e) there is a sedation before the tonification).

To sum up:

If the heart is stimulated at point H9 in the afternoon, the following results may be expected:

(a) The heart will be tonified, (if it was not in excess).
(b) The lung will be sedated, (if it was not deficient).
(c) The gall bladder will be sedated, (if it was in excess and the treatment was performed in the afternoon).
(d) The vessel of conception is sedated, (if in excess).
(e) The spleen and small intestine will be tonified, (if they were deficient).
(f) The liver and pericardium will be tonified, (if they were deficient)

These results are in conformity with the following acupuncture laws.

(a) Direct.
(b) The law of 'Husband-wife'.
(c) The law of 'Midday-midnight'.
(d) Special secondary meridian connections.
(e) Superficial circulation of energy.
(f) Deep circulation of energy (five elements).

Case History. A patient was seen who was suffering from heartburn and constipation. She had had typhoid fever forty years previously from which she nearly died. Since that time she has never felt well and lacked energy.

Pulse diagnosis revealed, amongst other things, a weakness of the pulses of the liver and large intestine. The liver and large intestine were tonified at their points of tonification—Liv8 and Li11. These points in

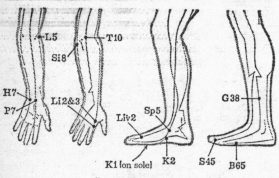

FIG. 41 Points of sedation

conjunction with other points, repeated over a considerable period, because of the chronicity of the disease, brought the patient back to a nearly normal state of health.

Points of Sedation

Each of the twelve main meridians has in contrast to its point of tonification, a specific point of sedation. This is invariably located on the meridian whose function it controls.

The point of sedation of a meridian is the 'son' point of its own element i.e. the small intestine belongs to the element fire. The 'son' (following element) of fire is earth. Therefore the point of sedation is the earth point—small intestine 8 (see chapter VI). The fifth column in chapter VI, section 'treatment via law of five elements', likewise gives the point of sedation of all meridians as also printed below (Fig. 41).

Meridian	*Point*
Lungs	point L5
Large intestine	point Li2 and Li3
Stomach	point S45
Spleen	point Sp5
Heart	point H7 (also source)
Small intestine	point Si8
Bladder	point B65
Kidney	point K1 and K2
Pericardium	point P7 (also source)
Triple warmer	point T10
Gall bladder	point G38
Liver	point Liv2

The effects are theoretically the reverse of the points of tonification, though this in practice will not always be found to operate, so that progress should be carefully followed by palpation of the pulse. It can sometimes happen that a point of tonification acts as if it were a point of sedation and vice versa.

Theoretically the effects expected should be:

(*a*) Direct stimulation—sedation.
(*b*) Stimulation in accordance with 'Husband-wife' law—tonification.
(*c*) Stimulation in accordance with the 'Midday-midnight' law (Yang organ in the morning, Yin organ in the evening)—tonification.
(*d*) Stimulation of special meridians—tonification.
(*e*) Stimulation in accordance with superficial circulation of energy —sedation.
(*f*) Stimulation in accordance with deep circulation of energy (five elements)—sedation.

Thus, if the liver were sedated at its point of sedation, point Liv2, the following would result:

(*a*) Sedation of liver, (if it were not already deficient).
(*b*) Tonification of spleen, (if it had been deficient before).
(*c*) Tonification of the small intestine (if Liv2 is sedated in the evening).
(*d*) Nil.

(e) Sedation of the gall bladder and lungs, (if these were in excess before treatment).

(f) Sedation of the kidney and heart, (if these were in excess before treatment).

Case History. The patient was lethargic in the morning, had frontal headaches, and palpitations with physical or even slight mental strain. Pulse diagnosis showed an over-activity of the gall bladder and an under-activity of the heart. The point of sedation of the gall bladder (G38) was stimulated in the morning which directly sedated the gall bladder; and at the same time indirectly tonified the heart via the law of midday-midnight (c). Her symptoms disappeared in ten minutes.

The Source

Each of the twelve main meridians has a third type of directive point called the source, which is located on the meridian that it controls. These are as follows (Fig. 42):

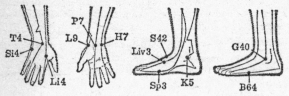

FIG. 42 Source points

Meridian	Point
Lungs	point L9 (also point of tonification)
Large intestine	point Li4
Stomach	point S42
Spleen	point Sp3
Heart	point H7 (also point of sedation)
Small intestine	point Si3
Bladder	point B64
Kidney	point K5
Pericardium	point P7 (also point of sedation)
Triple warmer	point T4
Gall bladder	point G40
Liver	point Liv3

The stimulation of the source point gives various results:

1. It may either cause direct tonification or direct sedation of the meridian on which it is placed. This type of amphoteric action distinguishes the source from a point of tonification or sedation which, as a rule, can only be tonified or sedated, as the name implies.

According to classical acupuncture the source is tonified if a gold needle is used, and is rotated clockwise, the needle being inserted in the direction of the current of energy along the meridian and the operation effected while the patient exhales. Similarly, the same point is sedated if a silver needle is used, if it is rotated anti-clockwise, and the needle is inserted against the direction of the current of energy along the meridian and the operation effected while the patient inhales.

In my experience the above factors are not of any great consequence. In fact in whatever way the source is stimulated it has the desired effect of re-establishing the balance of energy. If, for example, the kidney were underactive and the source of the kidney, point K5, were stimulated with either a silver or a gold needle, the kidney would be tonified. Once this direct effect of either tonification or sedation has taken place, the same interactions follow as for the points of tonification or sedation, following the same laws and conditions which governed their operations in the former instances, i.e. those of:

'Husband-wife'.
'Midday-midnight'.
Special secondary vessels.
Superficial circulation of energy.
Deep circulation of energy (five elements).

2. Stimulation of the source usually has a rapid effect.

3. If the point of tonification, or the point of sedation has been used and thereafter the source is utilised, the effect of the tonification or sedation is accentuated. e.g. If the kidney has been tonified by point K7, (its point of tonification) but the result is not sufficient, the source K5 if used, may reinforce the action.

Case History. A patient who had had polio as a child in one of his legs developed severe sciatica when an adult. Pulse diagnosis showed a weakness of the kidney pulse. This was tonified by putting a needle into the source of the kidney (point K5), which resulted in an immediate relief of pain, but only to return when he started walking again. Various other points were tried; the pain being relieved each time, only to re-appear again. It was then found that one leg was an inch shorter than the other. (The leg which had been affected by polio growing more slowly). A raised shoe relieved the condition. This type of gross struc-tural damage is not suitable for treatment by acupuncture.

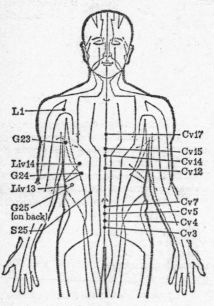

FIG. 43 Alarm points

Alarm Points

The points known as 'alarm points' are a series that occur on the ventral surface of the abdomen or chest. They are (Fig. 43):

Meridian	Point
Lungs	point L1
Large intestine	point S25

Stomach	point Cv12
Spleen	point Liv13
Heart	point Cv14
Small intestine	point Cv4
Bladder	point Cv3
Kidney	point G25
Pericardium	point Cv15 (Discovered by Soulié de Morant)
Triple warmer (main)	point Cv5
Triple warmer (superior)	point Cv17
Triple warmer (middle)	point Cv12
Triple warmer (inferior)	point Cv7
Gall bladder (main)	point G24
Gall bladder (secondary)	point G23
Liver	point Liv14

All these points are located on the embryological anterior surface of the body. Only three points are on the meridian which they subserve, viz: Lung, gall bladder and liver, these three being organs which follow one another in the superficial circulation of energy, occupying the time from 11 p.m. to 5 a.m. Many of the points of alarm are located on a meridian the vessel of conception (Cv) which does not belong to the primary system of twelve meridians.

The alarm points have various functions:

1. The points are all situated on the ventral surface, and this being a Yin surface, they are typically associated with diseases of a Yin type. This is so marked that in old textbooks, only the five primary Yin organs (liver, heart, spleen, lung and kidney) are described as having an alarm point.

To quote Zhenjiu Yixue—'*The illnesses of the Yang act on the Yin. That is why the points of alarm are all in the Yin. The front of the abdomen and chest are Yin; that is why the points of alarm are there.*'

Yin diseases are those which are typically accompanied by cold, depression and weakness.

2. In the characteristic type of Yin disease the point of alarm becomes excessively tender. e.g. In many cardiac diseases the point of alarm of the heart (Cv14) (Fig. 43) which is about 1 inch below the xiphoid process of the sternum, is spontaneously tender.

I have mentioned earlier that an acupuncture point that needs treatment often becomes spontaneously tender. This tenderness is so exaggerated in the case of the points of alarm that it is used as a palpatory method of diagnosis in the following manner.

The patient is asked to lie flat and relaxed on a couch, with the chest and abdomen bare. The points of alarm are then palpated and if they are more tender than the surrounding tissues a functional disturbance of the organ which they represent may be deduced.

The area of tenderness and of superficial tissue changes as shown by palpation is considerably larger and more easily noticeable in the case of the alarm points than in that of the other types of acupuncture points when comparably activated. These two factors, taken together with the relatively greater increase in tenderness of this type of point, constitute useful diagnostic criteria.

The alarm point may become spontaneously painful, so that the patient is aware of it without it being pressed, more easily than in any other type of activated acupuncture point. Naturally this makes diagnosis easier.

3. Normally the point of alarm is considered a point of tonification, which, if stimulated, increases the energy in the meridian which it subserves.

The tonification of the meridian-organ concerned is followed to some extent by a tonification of the meridian which preceeds and follows it in the superficial circulation of energy and also the deep circulation of energy (five elements).

4. In my experience the point of alarm serves equally well as a point of sedation but care must be exercised in sedating an over-active alarm point as a hypertonification may unwittingly be the result, with an acute exacerbation of the condition being treated. This exacerbation can sometimes be avoided by stimulating the point of alarm for only a few seconds instead of the customary minutes.

5. Usually a qualitative increase in the Yang elements at the pulse will be noted. This is an uncertain response.

Illustration. In patients with diseases of the upper digestive organs, very frequently the point of alarm of the stomach, Cv12, (Fig. 43) becomes spontaneously tender. A needle put into this point may cause immediate relief of upper abdominal distension and nausea. The fundamental condition however will have to be treated by other points.

Associated Points

Qi Bo: *'If you press with your finger on these points, the pain of the corresponding organ is immediately relieved.'*

(Nei Jing, Ch. 51)

All meridians have an associated point on the back along the medial course of the bladder meridian on each side of the vertebral column. According to Qi Bo (Fig. 44):

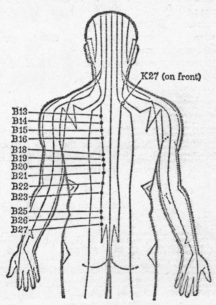

K27 (on front)

B13
B14
B15
B16
B18
B19
B20
B21
B22
B23
B25
B26
B27

FIG. 44 Associated points

Meridian	Point
Lung	point B13
Pericardium	point B14
Heart	point B15
(Governing vessel	point B16)

Liver	point B18
Gall bladder	point B19
Spleen	point B20
Stomach	point B21
Triple warmer	point B22
Kidney	point B23
Large intestine	point B25
Small intestine	point B27
Bladder	point B28

One special point to be noted is K27. This is considered to act as the associated point for the whole series.

The associated points, which are all paravertebral on the dorsal surface, have certain characteristics which are in contrast to the points of alarm:

1. Classically they are points of sedation. According to the laws of acupuncture, once the meridian concerned with a particular associated point is sedated, it in turn causes a sedation of the meridian which precedes it and the meridian which follows it, both in the superficial circulation of energy and in the deep circulation of energy. Classically, the procedure is the reverse of that which operates in the case of the points of alarm.

2. In my experience the associated points may be used with excellent results as points of tonification.

e.g. Point B23 is usually very efficacious in cases of under-activity of the kidney.

Although the point of alarm may cause an acute exacerbation if used in the inverse sense to that accorded by classical theory, this is not the case with the associated point.

3. These points have a general calming effect and are therefore used in Yang diseases such as over excitation and fever.

Li Kao Tong-iuann of the twelfth century writes:

'To treat a disease caused by wind or cold, you must stimulate the associated point of a storage, hollow organ. In fact the illness entered by the Yang and then flowed through the meridians. If it started by a cold exterior it must finish by returning to the exterior by warmth.'

4. Chinese osteopathy uses these points to correct small displacements of vertebrae. The rationale is as follows:

In a disease of the descending colon, the associated point of the large intestine, B25 (Fig. 45), on the same side as the descending colon, i.e. the left, will together with other points become spontaneously tender. This causes a spasm of the muscles in the vicinity of point B25 on the left. These muscles which are adjacent to, and attached to, the fourth lumbar vertebra, cause it to be displaced towards the left.

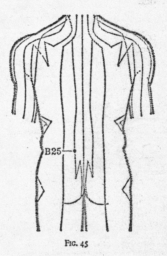

FIG. 45

Hence a disease of the descending colon may, under the correct conditions, cause as a secondary result a displacement of the fourth lumbar vertebra to the left, with, if the displacement is severe enough, resultant lumbago and possibly sciatica.

Only rarely does an internal disease cause a displaced vertebra for:

(a) Not all internal diseases cause tenderness of the associated point, and hence muscle spasm.

(b) The muscle spasm must be of a fairly severe degree.

(c) Before displacement occurs there must in general be associated factors which could operate to facilitate the displacement of a vertebra, such as a general metabolic disturbance causing osteo-porosis, or weakness of the paravertebral muscles, trauma, etc.

If the displacement of the fourth lumbar vertebra is only small it may be corrected by an acupuncture point (or medicine) that corrects the disease of the descending colon. It may also be cor-rected by stimulating point B25 on the left, though usually treat-ment of various secondarily effected points is also required.

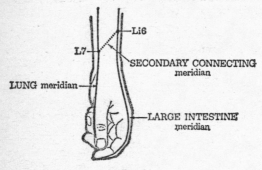

FIG. 46

A severe displacement can only be corrected by osteopathic manoeuvres or by manipulation under an anaesthetic. If the internal factor which, together with other factors, originally caused the displacement of the vertebra, are not treated at the same time as the displacement is corrected, there is a greater likelihood of a recurrence. This explains the too frequent recurrence of lumbago or sciatica if treated only by manipulation, osteopathy, corsets, etc. Conversely, under suitable conditions, the correction of the dis-placed vertebra may cure the primary internal disease.

This problem can, at times, present itself as the classical dilemma —which came first, the chicken or the egg? It is sometimes best to attack both ends of the problem at the same time.

Manipulative surgeons, osteopaths and masseurs, have often found that by manipulating vertebrae they can cure, or alleviate, internal diseases. This has been partially explained via neural reflexes connecting the diseased organ with the appropriate spinal segment. As all the important internal organs, (from the acupuncture point of view), have an associated point which is paravertebral, I think we can regard this fact as at least a partial explanation of the connection.

Illustration. Stage fright is sometimes due to an over-activity of the heart. The heart may be sedated in many ways; but as stage fright is an illness associated with nervousness the associated point is probably the best heart point to choose—point B15—between the shoulder blades.

'Connecting' Points

The so called 'connecting' points connect coupled meridians by a secondary meridian. e.g. There is a secondary meridian running from L7 to Li6, joining the lung meridian with the large intestine meridian (Fig. 46).

Coupled meridians are meridians which follow one another in the superficial circulation of energy and, at the same time, are of opposite sign, the one being Yin, the other Yang. It follows that the former lies on an embryological anterior surface, while the latter lies on an embryological posterior surface.

These coupled meridians thus constitute a unit of similarities and dissimilarities (Fig. 47).

Coupled meridians	Lungs	Yin	point L7
	Large intestine	Yang	point Li6
Coupled meridians	Stomach		point S40
	Spleen	Yin	point Sp4
Coupled meridians	Heart		point H5
	Small intestine	Yang	point Si7
Coupled meridians	Bladder		point B58
	Kidney	Yin	point K6
Coupled meridians	Pericardium		point P6
	Triple warmer	Yang	point T5
Coupled meridians	Gall bladder		point G37
	Liver	Yin	point Liv5

The vessel of conception, being the most Yin of any meridian and the governing vessel being the most Yang of any meridian, are also connected by their connecting points, these being Cv15 and Gv1 respectively.

There is also the so called Great Connecting Point of the spleen, Sp21.

Treatment of the connecting points can serve various purposes:

1. A disequilibrium between the two meridians of a couple may be corrected by only using one point. In this case either the connecting point of the deficient meridian is tonified or the connecting point of the over-active meridian is sedated.

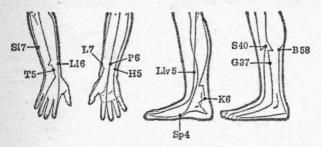

FIG. 47 Connecting points

If, for example, the liver is deficient, while the gall bladder is in excess, either the liver may be tonified at its first connecting point Liv5, or the gall bladder may be sedated at its connecting point G37. Thus two meridians are corrected by only using one acupuncture point.

It seems as if the connecting point on either meridian acts as a sort of short-circuit enabling the excess of energy to flow along the connecting vessel from one meridian to the other of a couple.

2. The connecting point controls the energy between the left and right halves of a meridian. e.g. If the left lung meridian has an excess of energy, while the right lung meridian is deficient in energy, these may be equilibriated by tonifying the connecting point of the lung meridian (L7) on the right side, or by sedating L7 on the left side— only one needle being used to produce the two effects desired.

3. The connecting point controls the flow of energy between organs related to one another by the law of 'midday-midnight'. If the bladder meridian has an excess of energy while the lung is deficient, (bladder maximal activity 4 p.m., lung maximal activity

4 a.m.) the connecting point L7 of the lung may be tonified, and thus equilibrium between bladder and lung may be achieved.

When the connecting point is used in obedience to the law of 'midday-midnight', account need not be taken of Yin or Yang times of day, the reaction taking place equally well whatever time the poiot is stimulated.

If an exchange of energy is desired between the left lung and left functional part of the bladder, only the connecting point on the left side is utilised.

Case History. A patient had chronic bladder trouble over a period of twenty years, with frequency, nocturia, burning pain on urination, etc. The pulse diagnosis showed an irritable bladder with a wiry pulse. Treatment of the bladder meridian did not alter the condition for, as may be noticed frequently in acupuncture, the direct treatment of the diseased meridian, is often of no value.   The bladder and kidney meridians are united via their connecting points—B58 and K6.   Stimulation of the kidney connecting point, K6, regularised the pulse of the bladder.   The treatment had to be repeated many times before a considerable improvement, though not complete cure, ensued.

GRAND PIQURE

By this method energy is drawn from one side of the body to the other via the connecting points.

If there is, for example, an excess of energy in the right stomach meridian, causing pain along part of its course, the connecting point on the opposite (left) side is stimulated—point stomach 40 (S40).

GRAND PIQURE COMBINED WITH TREATMENT ACCORDING TO THE LAW 'MIDDAY-MIDNIGHT'

The effect of the above example may be increased by stimulating the connecting point on the left side of the meridian connected to the stomach meridian by the law of 'midday-midnight'. This is the meridian of pericardium. Therefore its connecting point pericardium (P6) is stimulated only on the left.

Point of Entry and Exit

As mentioned previously, the energy Qi flows through the twelve meridians in a certain invariable sequence: lung, large intestine, stomach, spleen, etc.

This flow of Qi is always in the same direction starting at the point of entry of the lung meridian (Fig. 48), flowing along the lung meridian, and leaving again at the point of exit to enter the point of entry of the meridian of the large intestine; flowing along the meridian of the large intestine, to leave it again at the point of exit, to enter the meridian of the stomach at its point of entry etc., etc.

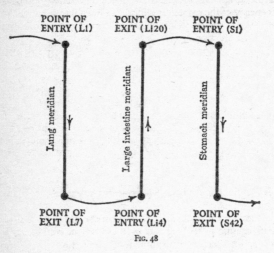

FIG. 48

The point of entry and the point of exit are usually the first and last points respectively of the meridian concerned, in which case, a secondary meridian unites the end of one meridian with the beginning of the next, along which the Qi flows. In some cases the point of entry or exit is not the end point of a meridian (though it is not far from it). In this case the main secondary meridian uniting the two meridians does not occur at the end point of the meridians. Nevertheless, the remaining distal portion of the meridian is not a cul-de-sac for the secondary meridian mentioned in the previous paragraph, uniting the end points of the meridians, is still operative, though it performs a function in this case secondary to the meridian connecting the points of exit and entry.

Tonification of a point of entry, tonifies the meridian concerned, provided the previous meridian has an excess of energy to pass on.

Sedation of a point of entry, under the reverse circumstances, is considered to sedate the meridian concerned. In my experience this effect is unreliable and unpredictable.

Sedation of a point of exit, sedates the meridian concerned, provided the following meridian is deficient in energy so that the excess energy of the sedated meridian may pass into it.

Tonification of the point of exit is considered to produce the same result as sedation.

The points of entry are more reliable in their effects than the points of exit (Fig. 49).

Points of Entry

Lung	point L1 (also point of alarm)
Large intestine	point Li4 (also source. *1st point is* Li1)
Stomach	point S1
Spleen	point Sp1
Heart	point H1
Small intestine	point Si1
Bladder	point B1
Kidney	point K1 (also point of sedation)
Pericardium	point P1
Triple warmer	point T1
Gall bladder	point G1
Liver	point Liv1

Points of Exit

Lung	point L7 (also 'Lo' point. *Last point is* L11)
Large intestine	point Li20
Stomach	point S42 (also point of tonification. *Last point is* S45)
Spleen	point Sp21
Heart	point H9 (also point of tonification)
Small intestine	point Si19
Bladder	point B67 (also point of tonification)
Kidney	point K22 (*Last point* K27)
Pericardium	point P8 (Alarm point of pericardium. *Last point* P9)

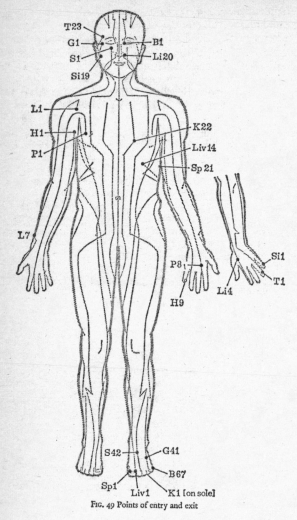

FIG. 49 Points of entry and exit

Triple warmer   point T23
Gall bladder   point G41 (*Last point* G44)
Liver   point Liv14 (Also point of alarm)

Example. Tonification of the point of entry of the small intestine
(Si1) will tonify the small intestine, provided the meridian of the
heart has an excess of energy to pass on.

Sedation of the point of exit of the triple warmer (T23) will
sedate the triple warmer provided the gall bladder is deficient in
energy.

Case History. In the skin disease, acne rosacea, the pulse of the large
intestine is, amongst others, weakened. The point of entry of the large
intestine, Li4, together with other points, were stimulated in a patient.
Within a month the condition was about 90% cured, despite the fact
that she had had this condition for many years.

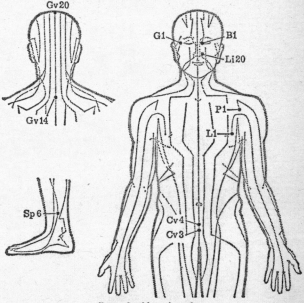

FIG. 50 Special meeting points

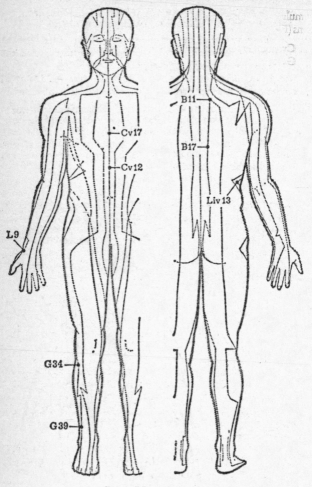

FIG. 51 Eight meeting points

Special Meeting Points

Stimulation of these points has an effect on a related group of meridians (Fig. 50):

Cv3 and Cv4	3 leg Yin and conception vessel
Gv20	3 leg Yang and governing vessel
Sp6	3 leg Yin
Gv14	7 Yang

The meeting point of the other related meridians is as follows:

Lung and Spleen	L1
Pericardium	P1
Small intestine and bladder	B1
Triple warmer and gall bladder	G1
Large intestine and stomach	Li20

Case History. A patient at hospital was suffering from a catarrhal nasal condition with hyperacidity. Pulse diagnosis showed a weak pulse of the large intestine and a bumpy wiry pulse of the stomach. The meeting point of the large intestine and stomach, point Li20, in combination with a more fundamental realignment of his fundamental energetics, cured the patient.

The Eight Meeting Points

The following eight points have a particular influence on the eight tissues mentioned; or, as the Chinese put it, 'the Qi of these eight tissues meet at the eight points.' They are (Fig. 51):

Solid organs (Zang)	Liv13
Hollow organs (Fu)	Cv12
Energy and/or breath (Qi)	Cv17
Blood	B17
Bones	B11
Marrow	G39
Muscles	G34
Vessels	L9

The Combining Points

The Nei Jing describes these points as:

'At the level of the combining points the energy of the six Yang

Fu penetrates into the interior of the body'. Again: 'The combining points rule the energies of the meridians.'

These points are (Fig. 52):

	Organ	Lower combining point	Upper combining point
3 Yang of foot	Stomach	S36	—
	Gall bladder	G34	—
	Bladder	B54	—
3 Yang of hand	Large intestine	S37	Li11
	Triple warmer	B53	T10
	Small intestine	S39	Si8

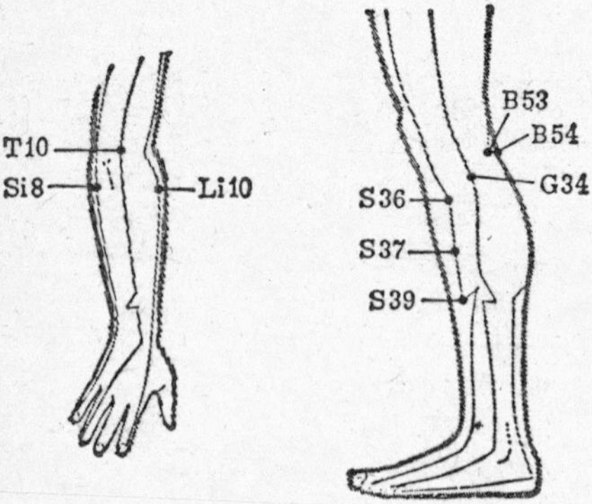

FIG. 52 Combining points

Frequently an organ-meridian may be disturbed in such a way that the disturbance is not seen at the extremities, but centrally in the abdomen and thorax. In these cases the combining points may be used.

Case History. A patient complained of intermittent swelling of the lower abdomen with no clear urinary symptoms. Pulse diagnosis revealed a wiry pulse of the bladder. The combining point of the bladder, point B54, in combination with subsidiary points, cured the condition.

The Points 'Window of the Sky'

The Nei Jing says:

'All the energies Yang come from the Yin, for the Yin is earth. This Yang energy always climbs from the lower part of the body towards the head; but if it is interrupted in its course it cannot climb beyond the abdomen. In that case one must find which meridian is diseased. One must tonify the Yin (as it creates the Yang) and disperse the Yang so that the energy is attracted towards the top of the body and the circulation is re-established.'

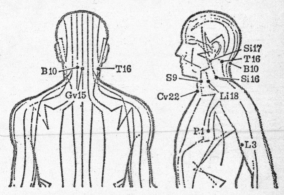

FIG. 53 Window of the sky points

The points used for this purpose are (Fig. 53):

S9	Li18
T16	B10
L3	Cv22
Gv15	Si16
Si17	P1

It will be noticed that all these points except for L3 and P1 are in the neck, which is the route whereby energy goes from the lower part of the body to the head.

The symptomatology according to the Nei Jing is as follows:
S9. Severe pain in the head, fullness of the chest, dyspnoea.
Li18. Loss of voice.

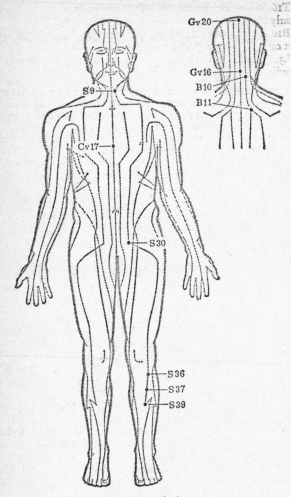

FIG. 54 'Four seas' points

T16. When the patient suddenly becomes deaf or cannot see clearly.

B10. Spasms, muscular contractions, fainting, when the patient's feet can no longer support the weight of the body.

L3. Great thirst (disharmony of liver and lung) nose bleeds or bleeding by the mouth.

Case History. A patient at hospital had lost his voice some months previously and felt light-headed. He felt as if his head and body were not properly connected. The point 'window of the sky' S9, repeated several times, in combination with subsidiary points, cured the patient. The first treatment only point S9 was used and caused a fluctuation between improvement and worsening of the condition. Thereafter (as the barrier had been opened) the appropriate Yin meridians (liver and kidney in this case) were tonified and the light-headedness disappeared.

The Points of the 'Four Seas'

The Nei Jing says:

'*Man possesses four seas and twelve meridians, which are like rivers that flow into the sea.*'

The four seas are:

1. The sea of nourishment.
2. The sea of blood.
3. The sea of energy.
4. The sea of the bone marrow.

1. 'The sea of nourishment is represented by the stomach. Its two principal points are S30 and S36 (Fig. 54).

If there is excess the patient has abdominal swelling. If there is emptiness he cannot eat.'

2. 'The sea of the blood is represented by the extra meridian, the 'penetrating vessel', which is the sea of the twelve meridians. Its points of liaison are B11 for the high part of the body and S37 and S39 for the low part of the body.

When there is an excess, the patient has the sensation that his body is greater in volume. When there is deficiency, the patient is affected, but cannot define what he feels.'

3. 'The sea of energy is represented by the region around the point Cv17. These points are in liaison with point B10 behind and S9 in front of the neck.

If there is excess, the patient feels pain in the chest, the face is red and there is breathlessness. If there is emptiness the patient cannot speak.'

4. 'The sea of bone marrow. Its point of liaison is localised on the summit of the head (probably governing vessel 20) and at the back of the head by point governing vessel 16.

If there is fullness, the patient feels as if he has an excess of energy.

If there is emptiness, the patient has dizzy bouts, noises in his ears, fainting, pain in the calf.'

It says further:

To sum up: 'One must be able to discern accurately if there is emptiness or fullness and puncture the points of the 'four seas' correctly, for thus one can regularise all the energies. But if one punctures incorrectly one can provoke grave trouble.'

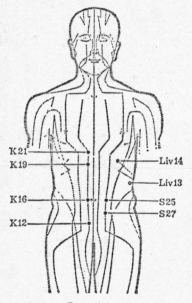

FIG. 55 Shokanten

Shokanten

The Japanese Shokanten, as described by Manaka*, are points on the abdomen that become tender if the greater, lesser or middle Yang or Yin become affected. As such, these points may be used both diagnostically and in the reverse, therapeutically (Fig. 55).

Greater Yang	Small intestine/Bladder—K12.
Lesser Yang.	Triple warmer/Gall bladder—S25.
Sunlight Yang	Large intestine/Stomach—S27.
Greater Yin	Lung/Spleen—Liv13.
Absolute Yin	Pericardium/Liver—Liv14.
Lesser Yin	Heart/Kidney—K16.

In addition, Manaka has described:

Lesser Yang	K21.
Absolute Yin	K19.

The Accumulating Points (Hung)

The great gilded Buddhist temples in China are known as the 'Hung' and the name is also used for the accumulating points, for they are as important in the body as temples. They are described as 'gaps in the body where Qi and blood converge and collect' and are used in chronic disease (Fig. 56).

Meridian	Accumulating point
Lung	L6
Large intestine	Li7
Stomach	S34
Spleen	Sp8
Heart	H6
Small intestine	Si6
Bladder	B63
Kidney	K4
Pericardium	P4
Triple warmer	T7
Gall bladder	G36
Liver	Liv6

*IV Journees Internationales d'Acupuncture.

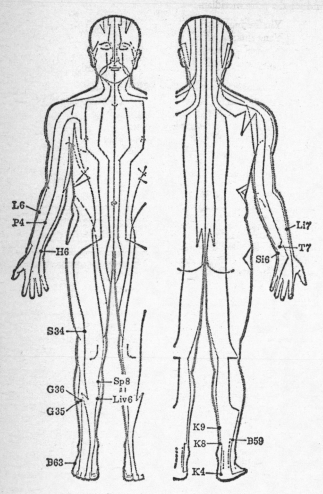

FIG. 56 Accumulating points

And for the extra meridians:—

Yin linking vessel	K9
Yang linking vessel	G35
Yin heel vessel	K8
Yang heel vessel	B59

The Five Categories

The Chinese give special names to five (or six) categories of points.

Category	Chinese name	Meaning
I	Well	Emerging
II	Gushing	Flowing
III	Transporting	Pouring
IV	Penetrating	Moving
V	Uniting	Entering

These names, and the meaning of them to the Chinese, image the

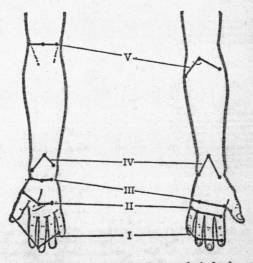

FIG. 57 Five categories. The acupuncture points are the same as for the five elements (Fig. 37)

flow of Qi along the meridian as the movement of water. The 'well' is the place whence the water emerges; the 'gushing' is its overflow; the 'transporting' point is where the water pours along; the 'penetrating' point where it moves. Finally, since water must find its way to union with the sea, the solid and hollow organs, we have the 'uniting point'.

The categories I, II, III, IV and V follow the points of the five elements: (for the Yin meridians) wood, fire, earth, metal, water; (for the Yang meridians) metal, water, wood, fire, earth, in that order from the tips of the fingers or toes to the elbow or knee (Fig. 57 and 58).

	CATEGORY*				
	I	II	III	IV	V
Lungs	L11	L10	L9	L8	L5
Spleen	Sp1	Sp2	Sp3	Sp5	Sp9
Heart	H9	H8	H7	H4	H3
Kidney	K1	K2	K5	K7	K10
Pericardium	P9	P8	P7	P5	P3
Liver	Liv1	Liv2	Liv3	Liv4	Liv8
Large intestine	Li1	Li2	Li3	Li5	Li11
Stomach	S45	S44	S43	S41	S36
Small intestine	Si1	Si2	Si3	Si5	Si8
Bladder	B67	B66	B65	B60	B54
Triple warmer	T1	T2	T3	T6	T10
Gall bladder	G44	G43	G41	G38	G34
Yin element	Wood	Fire	Earth	Metal	Water
Yang element	Metal	Water	Wood	Fire	Earth

*The system of classifying the categories adopted by Soulié de Morant and after him, in the first edition of this and my other books, utilised six categories; the additional category, IV, being the source points mentioned earlier in this chapter. Thus, under that system category III and IV were exactly the same point for the Yin meridians, but different for the Yang meridians. In both systems categories I, II and III are the same, whilst V and VI have become IV and V in this book. The five category system used in the 2nd edition of this book is the one more generally used in China today. I will therefore at a later date, alter the six category system used in the present editions of my other books, to the five category system.

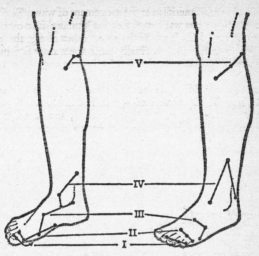

FIG. 58 Five categories. The acupuncture points are the same as for the five elements (Fig. 38)

Traditionally the following groups of diseases can best be treated by them:

I   Region below heart full.
II Body hot.
III Body heavy, joints painful.
IV Dyspnoea, coughing, cold and hot.
V   Rebellious Qi and diarrhoea.

Meeting Points

The meeting points have an effect on more than one meridian. For example bladder 1, as can be seen from the table, influences the small intestine and stomach in addition to the bladder itself.

It is difficult, if not impossible, to tell if a certain meeting point exerts its effect on several meridians directly, or indirectly via the normal laws of acupuncture. If the latter idea were correct, it should, of course apply to all points on the same meridian—in which case the conception of meeting points would be superfluous.

For the above reasons it is difficult to tell which are the various meridians effected by a meeting point, and opinions vary. Five traditional sources are given, which the reader may compare.

Likewise the meeting points given in my other books, are essentially but not quite the same, as in this volume.

Meeting Points

	1	2	3	4	5
L1	L	L, Sp	Sp	L, Sp	L, Sp
Li14	Li con	Li con	Li con	Si, B, Ya-l	Li con, Si, B, Ya-l
Li15	Li, Ya-h	Li, Ya-h	Li, Ya-h	Li, Ya-h	Si, Li, Ya-h, G
Li16	Li, Ya-h	Li, Ya-h	Li, Ya-h	Li, Ya-h	Li, Ya-h
Li20	Li, S	Li, S	Li, S	Li, S	Li, S
S1	Ya-h, Cv, S	Ya-h, Cv, S	Ya-h, Cv, S	Ya-h, Cv, S	Ya-h, Cv, S
S3	Ya-h, S	Ya-h, S	Ya-h, S	Li, S, Ya-h	Ya-h, S
S4	Ya-h, Li, S	Ya-h, Li, S	Ya-h, Li, S	Ya-h, Li, S	Ya-h, Li, S, Cv
S7	S, G		S, G	S, G	S, G
S8	G, Ya-l		G, S	G, S	G, S
S9				S, G	S, G
S30				Pen	Pen
Sp6	Sp, Liv, K	Sp, Liv, K	Sp, Liv, K	Sp, Liv, K	Sp, Liv, K
Sp12	Sp, Liv	Sp, Yi-l	Sp, Liv		Sp, Liv
Sp13	Sp, Liv, Yi-l	Sp, Yi-l	Sp, Liv, Yi-l	Sp, Liv, Yi-l	Sp, Liv, Yi-l
Sp15	Sp, Yi-l	Sp, Yi-l	Sp, Yi-l	Sp, Yi-l	Sp, Yi-l
Sp16	Sp, Yi-l	Sp, Yi-l	Sp, Yi-l	Sp, Yi-l	Sp, Yi-l

Ya-l = Yang linking vessel
Yi-l = Yin linking vessel
Ya-h = Yang heel vessel
Yi-h = Yin heel vessel
Pen = Penetrating vessel
Gir = Girdle vessel
con = connecting meridian

Taken from:
1. Jia yi ying
2. Wai tai mi yao
3. Tong ren yu xue tu jing
4. Zhen jiu da cheng
5. Lei jing tu yi
After Jingluoxue Tushuo by Hiu-jan and Zhu Ru-gong

	1	2	3	4	5
Si10	Si, Ya-l, Ya-h	Si, B, Ya-l Ya-h	Si, B, Ya-l, Ya-h	Si, Ya-l, Ya-h	Si, B, Ya-l, Ya-h
Si12	Si, Li, T, G	Si, Li, T, G	Si, Li, T, G	Si, Li, T, G	Si, Li, T, G
Si18	Si, T	Si, T	Si, T	Si, T	Si, T
Si19	Si, T, G	Si, T, G	Si, T. G	Si, T, G	Si, T, G
B1	Si, B, S	Si, B, S, Li	Si, B, T, G, S	Si, B, S, Ya-h, Yi-h	Si, B, S
B11	B, Si	B, T	B, G	B, Si, G, T	Gv con, B, Si
B12	Gv, B	Gv, B	Gv, B		Gv, B
B33		Liv		Liv, G	
B36	B	B, Si	B, Si	B, Si	B, Si
B59	Ya-h	Ya-h	Ya-h	Ya-h	Ya-h
B61		B, Ya-h		Ya-h	B, Ya-h
B62	Ya-h	Ya-h	Ya-h	Ya-h	Ya-h
B63	Ya-l	Ya-l	Ya-l	Ya-l	Ya-l
K3	Yi-h	Yi-h	Yi-h	Yi-h	Yi-h
K8	Yi-h	Yi-h	Yi-h	Yi-h	Yi-h
K9	Yi-l			Yi-l	Yi-l
K11	Pen, K	Pen, K	Pen, K	Pen, K	Pen, K
K12	Pen, K	Pen, K	Pen, K	Pen, K	Pen, K
K13	Pen, K	Pen, K	Pen, K	Pen, K	Pen, K
K14	Pen, K	Pen, K	Pen, K	Pen, K	Pen, K
K15	Pen, K	Pen, K	Pen, K	Pen, K	Pen, K
K16	Pen, K	Pen, K	Pen, K	Pen, K	Pen, K
K17	Pen, K	Pen, K	Pen, K	Pen, K	Pen, K
K18	Pen, K	Pen, K	Pen, K	Pen, K	Pen, K
K19	Pen, K	Pen, K	Pen, K	Pen, K	Pen, K
K20	Pen, K	Pen, K	Pen, K	Pen, K	Pen, K
K21	Pen, K	Pen, K	Pen, K	Pen, K	Pen, K
P1	P, G	P, G	P, G	P, Liv, T, G	P, G
T13		Li con	Li con	T, Ya-l	Li, T
T15	T, Ya-l	G, Ya-l	T, Ya-l	T, G, Ya-l	T, G, Ya-l
T17	T, G	T, G	T, G	T, G	T, G
T20	T, G, Li		T, G	T, G, Si	T, G, Si
T22	T, G, Si	T, G		T, G, Si	T, G, Si
G1	T, G, Si	T, G	T, G, Si	T, G, Si	T, G, Si
G3	T, G, S		G, S	T, G, S, Li	T, G, S

	1	2	3	4	5
G4	T, S	G, S	T, G, S, Li	T, G, S, Li	T, G, S, Li
G5				T, G, S, Li	
G6	T, G, S, Li	T, G, S, Li	T, G, S, Li	T, G, S, Li	T, G, S, Li
G7	G, B	G, B		G, B	G, B
G8	G, B		G, B	G, B	G, B˙
G9				G, B	G, B
G10	G, B		G, B	G, B	G, B
G11	G, B	T, G, B, Si	G, B	T, G, B	G, B
G12	G, B	G, B	G, B	G, B	G, B
G13	G, Ya-l	G, Ya-l	G, Ya-l	G, Ya-l	G, Ya-l
G14	G, Ya-l		G, Ya-l	T, G, Li, S, Ya-l	G, Ya-l
G15	G, B, Ya-l	G, B	G, B	G, B, Ya-l	G, B, Ya-l
G16	G, Ya-l	G, Ya-l	G, Ya-l	G, Ya-l	G, Ya-l
G17	G, Ya-l	G, Ya-l	G, Ya-l	G, Ya-l	G, Ya-l
G18	G, Ya-l	G, Ya-l	G, Ya-l	G, Ya-l	G, Ya-l
G19	G, Ya-l	G, Ya-l	G, Ya-l	G, Ya-l	G, Ya-l
G20	G, Ya-l	G, Ya-l	G, Ya-l	G, Ya-l, T	G, Ya-l
G21	T, Ya-l	T, G, Ya-l	T, G, Ya-l	T, G, S, Ya-l, T	T, G, S, Ya-l
G24	G, Sp		G, Sp, Ya-l	G, Sp, Ya-l	G, Sp, Ya-l
G26	G, Gir			G, Gir	G, Gir
G27	G, Gir			G, Gir	G, Gir
G28	G, Gir	G, Gir	G, Gir	G, Gir	G, Gir
G29	G, Ya-h	G, Ya-h	G, Ya-h	G, Ya-h	G, Ya-h
G30	G, S			G, S	G, S
G36	Ya-l	Ya-l	Ya-l	Ya-l	Ya-l
Liv13	Liv, G	Liv, G	Liv, G	Liv, G	Liv, G
Liv14	Liv, Sp, Yi-l	Liv, Sp, Yi-l	Liv, Sp, Yi-l	Liv, Sp, Yi-l	Liv, Sp, Yi-l
Cv1	Cv, Gv, Pen	Cv, Gv, Pen	Cv, Gv, Pen	Cv, Gv, Pen	Cv, Gv, Pen
Cv2	Cv, Liv	Cv, Liv	Cv, Liv	Cv, Liv	Cv, Liv
Cv3	Cv, Liv, Sp, K	Cv, Liv, Sp, K	Cv, Liv, Sp, K	Cv, Liv, Sp, K	Cv, Liv, Sp, K
Cv4	Cv, Liv, Sp, K	Cv, Liv, Sp, K	Cv, Liv, Sp, K	Cv, Liv, Sp, K	Cv, Liv, Sp, K, S
Cv7	Cv, Pen	Cv, Pen, K		Cv, Pen, K	Cv, Pen, K
Cv10	Cv, Sp	Cv, Sp	Cv, Sp	Cv, Sp	Cv, Sp

	1	2	3	4	5
Cv12	Cv, Si, T, S	Cv, Si, T, S	Cv, Si, T, S	Cv, Si, T, S	Cv, Si, T, S
Cv13	Cv, S, Si	Cv, S, Si	Cv, S, Si	Cv, S, Si	Cv, S, Si
Cv17				Cv, B, K, Si, T	
Cv23	Cv, Yi-l	Cv, Yi-l	Cv, Yi-l	Cv, Yi-l	Cv, Yi-l
Cv24	Cv, S	Cv, S	Cv, S	Cv, Gv, S, Li	Cv, S
Gv1			K, G	K, G	K
Gv13	Gv, B	Gv, B	Gv, B	Gv, B	Gv, B
Gv14	6 Yang, Gv	6 Yang, Gv	6 Yang, Gv	6 Yang, Gv	6 Yang, Gv
Gv15	Gv, Ya-l	Gv, Ya-l	Gv, Ya-l	Gv, Ya-l	Gv, Ya-l
Gv16	Gv, Ya-l	Gv, Ya-l	Gv, Ya-l	Gv, Ya-l, B	Gv, Ya-l
Gv17	Gv, B	Gv, B	Gv, B	Gv, B	Gv, B
Gv20	Gv, B	Gv, B	Gv, B	Gv, 6 Yang	Gv, B, T, G, Liv
Gv26	Gv, Li, S	Gv, Li	Gv, Li,	Gv, Li, S	Gv, Li, S
Gv28				Gv, Cv, S	Gv, Cv

Origin and End Points

Ma Chen-tai's school of thought during the Ming dynasty, compares the meridians to a river. The origin point is the source of the river; the end point is the lake into which the waters of the river accumulate at the end of its course. These points are only given for the three lower Yin and Yang; the origin in each case being at the ends of the toes, the end points of the trunk or the face. (Fig. 59).

Meridian (river)	Origin	End
Bladder	B67	B1
Stomach	S45	S8
Gall bladder	G44	Si19
Spleen	Sp1	Cv12
Kidney	K1	Cv23
Liver	Liv1	Cv18

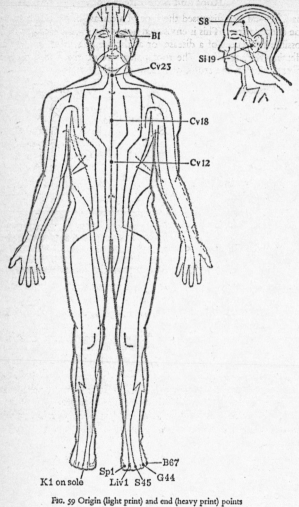

FIG. 59 Origin (light print) and end (heavy print) points

Root and Side Effect Points

Ma Chen-tai also discussed the root of a disease or meridian and its side effect or fruit. This is envisaged not as the main effect but as the possible side effect of a disease, or as the manifestation of a disease affecting the meridian. The root points are on the limbs, the side effect points on the trunk or head (Fig. 60).

Meridian	Root	Side Effect
Bladder	B59	B1
Gall bladder	G44 and G43	Si19
Stomach	S45	S9 and S4
Spleen	?Sp6	B20 and Cv23
Kidney	K8	B23
Liver	?Liv4	B18
Small intestine	Si6	B2
Triple warmer	T3	T23
Large intestine	Li11 and Li14	Li20
Lung	L9	L1
Heart	H7	B15
Pericardium	P6	P1

Selection of Acupuncture Points

Acupuncture points that are near the site of symptoms often have a greater local effect especially in painful conditions.

Points that are far away, especially the important points below the knee and elbow, often have a greater systemic effect.

Often the associated and alarm points stimulated together have a greater effect than each alone.

Some like to combine the source and connecting points.

The accumulation point and the appropriate one of the eight meeting points may be combined advantageously. For example S34 the accumulating point of the stomach may be used with Cv12 the meeting point of the Yang hollow organs (Fu) in epigastric pain with heartburn.

SUMMARY OF SOME OF THE MORE IMPORTANT TYPES OF ACUPUNCTURE POINT

	WOOD / METAL	FIRE / WATER	EARTH / WOOD	METAL / FIRE	WATER / EARTH	TONIFICATION	SEDATION	SOURCE	ASSOCIATED	ALARM	CONNECTING	ACCUMULATING	ENTRY	EXIT	LOWER MEETING PT.	UPPER MEETING PT.
Lung	L11	L10	L9	L8	L5	L9	L5	L9	B13	L1	L7	L6	L1	L7	……	……
Spleen	Sp1	Sp2	Sp3	Sp5	Sp9	Sp2	Sp5	Sp3	B20	Liv13	Sp4	Sp8	Sp1	Sp21	……	……
Heart	H9	H8	H7	H4	H3	H9	H7	H7	B15	Cv14	H5	H6	H1	H9	……	……
Kidney	K1	K2	K6	K7	K10	K7	K1	K6	B23	G25	K5	K4	K1	K22	……	……
Pericardium	P9	P8	P7	P5	P3	P9	P7	P7	B14	Cv17	P6	P4	P1	P8	……	……
Liver	Liv1	Liv2	Liv3	Liv4	Liv8	Liv8	Liv2	Liv3	B18	Liv14	Liv5	Liv6	Liv1	Liv14	……	……
Large intestine	Li1	Li2	Li3	Li5	Li11	Li11	Li2	Li4	B25	S25	Li6	Li7	Li4	Li10	S37	Li11
Stomach	S45	S44	S43	S41	S36	S41	S45	S42	B21	Cv12	S40	S34	S1	S42	S36	Si8
Small intestine	Si1	Si2	Si3	Si5	Si6	Si3	Si4	Si4	B27	Cv4	Si7	Si6	Si1	Si19	S39	Si8
Bladder	B67	B66	B65	B60	B54	B67	B65	B64	B28	Cv3	B58	B63	B1	B67	B54	S39
Triple warmer	T1	T2	T3	T6	T10	T3	T10	T4	B22	Cv5	T5	T7	T1	T23	B53	S8
Gall bladder	G44	G43	G41	G38	G34	G43	G38	G40	B19	G24	G37	G36	G1	G41	G34	T10

	MASTER	COUPLED	ACCUM-ULATING	CON-NECTING
Conception vessel	L7	K3		Cv15
Governing "	Si3	B62		Gv1
Penetrating "	Sp4	P6		
Girdle "	G41	T5		
Yin heel "	K3	B62	K8	
Yang heel "	B62	K3	B59	
Yin linking "	P6	Sp4	K9	
Yang linking "	T5	G41	G35	

THE FOUR SEAS

	MASTER	ACCUM-ULATING	CON-NECTING
Sea of nourishment	S30	S36	
Sea of blood	B11	S37	S39
Sea of energy	Cv17	B10	S9
Sea of marrow	Gv20	Gv16	

EIGHT MEETING POINTS

Zang	Liv13
Fu	Cv12
Qi	Cv17
Blood	B17
Bones	B11
Marrow	G39
Muscles	G34
Vessels	L9

MEETING POINTS

3 Lower Yang & Du mo	Gv20
3 Lower Yin	Sp6
7 Yang	Gv14
3 Lower Yin and Ren mo	Cv3 Cv4
Lung and spleen	L1
Pericardium and liver	P1
Small intestine and bladder	B1
Triple warmer and gall bladder	G1
Large intestine and stomach	Li20

WINDOW OF THE SKY

S9 Li18 Tr16 B10 L3 Gv16 Cv22 Si16 Si17 P1

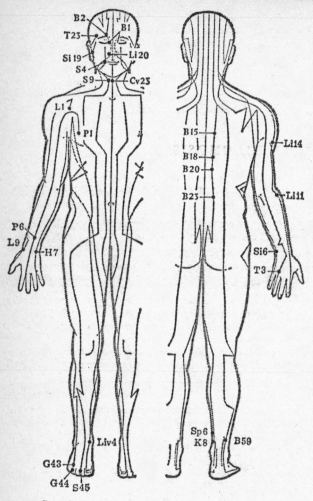

FIG. 60 Root (light print) and side effect (heavy print) points

IX

PULSE DIAGNOSIS

The pulse diagnosis is the key-stone of Chinese traditional diagnosis. It is described in detail in the ancient treatises (Fig. 61).*

'One should feel whether the pulse is in motion or whether it is still.'

'When the upper pulse is abundant, then the rebellious Qi rises. When the lower part is abundant, then the Qi causes a swelling in the abdomen. If the pulse appears to stop then the Qi has decayed.'

(Su Wen, Ch. 17)

'The "feon" pulse is like a weak wind that puffs up the feathers on the back of a bird, flustering and humming; like the wind that blows over autumn leaves; like water that moves the same swimming piece of wood up and down...

'If the pulse (at position III, left, deep) of the kidney is slightly hard... resistant... it is normal. If it is very hard, as hard as a stone, there will be death...'

(Hübbotter, P. 179)

A doctor skilled in its practice would—without ever speaking to or seeing the face or body of his patient and with no more contact than a hand thrust through a hole in a curtain to give access to the radial artery of the wrist—be able to arrive at a reliable diagnosis in a matter of minutes.

It can be used to confirm a diagnosis already arrived at by clinical

*Chinese pulse diagnosis, Sung dynasty. After Ilse Veith. The Yellow Emperor's Classic of Internal Medicine, 1949. Williams and Wilkins, Baltimore.

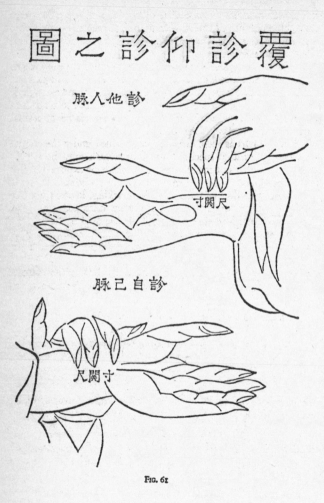

圖之診仰診覆

脉人他診

寸関尺

脉己自診

尺関寸

Fig. 61

and laboratory methods. It can be of very great benefit in a case where, although the patient is obviously ill, it has not been possible to arrive at a conclusive diagnosis in spite of thorough clinical and laboratory investigations.

It is so sensitive a method of diagnosis that not infrequently it will register past illnesses so accurately that a doctor is in a position to recount the past history of his patient's health (even though it involve illness he suffered fifty years previously) and to warn him of illness to be expected in the future, whether it be in several months' or in several years' time.

But such results as these can only be obtained under the correct conditions and within specific limitations, which must be strictly adhered to, so that not too much, nor too little, is expected of this method.

To those who do not understand its working, it can seem like magic. A patient who has been told by his doctor that at some time in the future he will develop a disease, though there is at the time no obvious indication to suggest it, may well conclude, when the 'prophecy' comes true, that his medical advisor has access to the secret mysteries of Nature.

To African Pygmies, utterly ignorant of the laws of aeronautics, an aeroplane taking flight like a giant bird above their heads, can

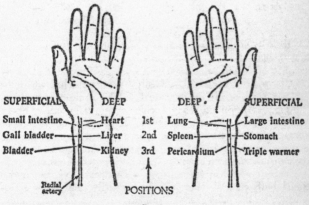

FIG. 62

only be explained in terms of magic. So we Europeans in our turn, observing some of the things the Pygmies can do and being ignorant of the processes which make them possible, either dismiss them casually from our minds or apply to them that same word of 'magic'. In reality, both conclusions are no more than misconceptions and both are due to insufficient knowledge.

The pulse diagnosis is both a science and an art. The scheme in Fig. 62 describes how pulse diagnosis works, though in actual practice it can only be satisfactorily learnt by demonstration and continuous correction from a doctor who has mastered the technique.

The pulse at the radial artery of the wrist is divided into three zones, each of which has a superficial and a deep position.

Left hand			Right hand	
Superficial	*Deep*		*Deep*	*Superficial*
		Positions		
Small intestine	Heart	1st	Lung	Large intestine
Gall bladder	Liver	2nd	Spleen	Stomach
Bladder	Kidney	3rd	Pericardium	Triple warmer

Each position occupies about half an inch of the radial artery— the exact amount can only be judged when one has become practised in this art, as it varies in each individual. The second position is roughly opposite to the radial apophysis.

If the ball of the finger is lightly placed on the radial artery in these three positions, it will be noticed, except in a perfectly healthy person, that the sensation is different at each place, and if gradually a greater pressure is exerted, suddenly a point is reached where the sensation has a totally different quality. This is the deep position. The superficial position has been compared to the elasticity of the arterial wall, and the deep position to the sensation of the flow of blood within the artery. It has been suggested that the pressure required for the superficial pulse diagnosis is the diastolic pressure, while that for the deep pulse diagnosis is the systolic pressure.

A patient who has, for example, a duodenal ulcer, as demonstrated both clinically and roentgenologically, will show a disturbance on the pulse marked 'stomach', i.e. the second position,

The standardised interpretation of the pulse given overleaf, is not shared by all authors, as the accompanying table shows.
Taken from:

		1	2	3	4	5	6	7	8
Left radial artery 1st	Deep	H	H	H	Sternum (Cvr7)	H	Sternum (Cvr17)	H	Sternum (Cvr17)
	Superficial	Si	Si	Si	Liv	H	Diaphragm	Pericardium connection	H
2nd	Deep	Liv	Liv	Liv	G	Liv	Liv	Liv	Liv
	Superficial	G	G	G	K	G	G	G	Si B
3rd	Deep	K	K	K	Si	K	K	K	Middle of chest L
	Superficial	B	B	B	L Middle of chest	B Li	Si B	L Middle of chest	S
Right radial artery 1st	Deep	L	L	L	S	L	L Middle of stomach	Sp	Sp
	Superficial	Li	Li	Li	Sp		S	S	Li
2nd	Deep	Sp	Sp	Sp	K	Sp	Sp	Gv4 T	K
	Superficial	S	S	S	Li	S	S	S	
3rd	Deep	Gv4	Gv4	T		K	K	Gv4 T	Li
	Superficial	B	T	Pericardium connection	Li	T Si	Li	Si	K

1. Wang shu-he 2. Li dong-yuan 3. Hua bo-ren 4. Li shi-zhen 5. Yu jia-yan 6. Li shi-cai 7. Zhang jing-yue 8. Yizong jinjian. Taken from Zhongyixue Gailun.

superficial, right. Similarly, as discussed later, other diseases show on other pulses, though not always with an obvious correlation.

There is some doubt as to the existence of both a superficial and a deep position in pulse diagnosis. As the superficial and deep position belong to coupled organs, the theory can in either case be easily adapted.

Case History. A girl with asthma missed so many of her lessons that discussions were in progress about sending her to a special school for the handicapped. From the point of view of acupuncture asthma may be due to a dysfunction of the lung or spleen or heart or kidney or liver or a non-specific type. Each one of the above six varieties has to be treated differently, and the pulse diagnosis is one of the best methods of differentiation. In this case the spleen was at fault, which when treated cured her asthma.

The Various Qualities

1. The pulse is different in every individual. There is no absolute norm, and something that may be normal for one individual, is pathological for another. Thus, the basic fundamental norm for each individual must be judged from experience, otherwise an attempt could be made to correct something which is normal so that disease would result.

It is, for example, perfectly normal for certain people to be vivacious and quick, and this is reflected in the quality of the radial pulse. For others, it is more normal to be of a phlegmatic temperament, and this is again reflected at the radial pulse. If an attempt be made to apply an artificial standard norm, to 'correct' these different (but in each case normal) pulses by 'appropriate' sedating or tonifying acupuncture points, disease will result. It is, after all, normal for an African to be black and a European to be white— an Albino African is ill.

2. All the pulses in ensemble may be plethoric, in which all the twelve basic pulses beat too strongly and feel over-full. This is known as 'total plethora of Yin and Yang'.

All the superficial pulses in ensemble may be plethoric. This is called 'total plethora of Yang'.

All the deep pulses in ensemble may be plethoric. This is 'total plethora of Yin'.

3. All the pulses in ensemble may be much too weak. This may be called 'total' weakness of Yin and Yang'.

Similarly we have:

'total weakness of Yang', and

'total weakness of Yin'.

4. If the pulses in position I are more powerful than the pulses in position III, Yang is more powerful than Yin. Similarly, if position III is stronger than position I, Yin is in excess of Yang.

5. If the pulses on the right radial artery are stronger than those on the left radial artery, there is excess of Yang. Conversely, if those on the left are stronger than those on the right, there is an excess of Yin.

Specific Qualities

Classically there are twenty-eight different qualities, though less than this is sufficient for ordinary practice.

'*When the pulse of the spleen is soft and even, well separated, as the footsteps of a chicken touching the ground; it is called regular. When the pulse is full, the frequency increases, like a chicken lifting its feet; then one speaks of disease.*'

(Jia Yi Jing, Vol. IV, Ch. 1a)

What might be called an artistic sense is a prerequisite for some of the finer points in pulse diagnosis, as frequently pulse conditions are felt which have not been felt before, or are not described in books (Fig. 63).

Essentially the pulse acquires the same quality (in an artistic sense) as the organ which it represents. For example, I once felt the pulse of a doctor who did not tell me his symptoms or the result of investigations. The pulse of his stomach was like thickened, wet, soggy, blotting paper. I was unable to think of the diagnosis (it obviously being not a stomach ulcer, carcinoma or hyperacidity). The doctor then told me that he had, as diagnosed gastroscopically, hypertrophic gastritis. The similarity (artistically speaking) between a hypertrophic gastric mucosa and thickened, wet, soggy, blotting paper are obvious. If my imagination had been a little livelier at the time, I am sure that I could have made the full diagnosis without the doctor concerned saying anything.

1. A particular quality developed on only one flank of the same radial artery in the same position, e.g. a plethoric heart pulse which

is more marked on the left side (lateral), suggests (to those who think they are differentiates) that the left side of the heart is more strained than the right as is usual in hypertension.

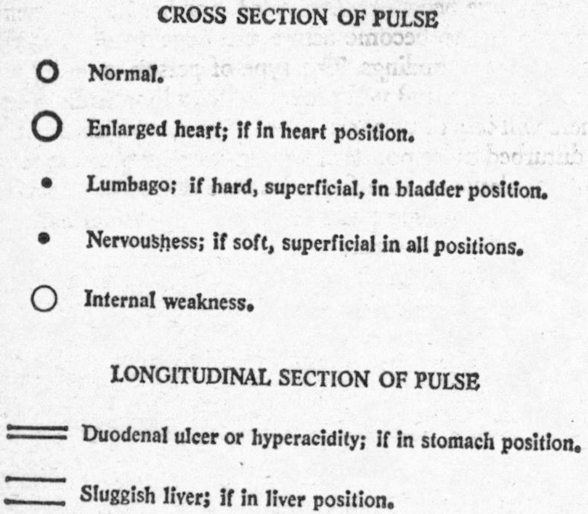

CROSS SECTION OF PULSE

O Normal.

O Enlarged heart; if in heart position.

• Lumbago; if hard, superficial, in bladder position.

• Nervousness; if soft, superficial in all positions.

O Internal weakness.

LONGITUDINAL SECTION OF PULSE

≡ Duodenal ulcer or hyperacidity; if in stomach position.

— Sluggish liver; if in liver position.

FIG. 63

2. The disturbance in the proximal or distal part of the pulse is more marked. For example, in the case of the pulse of the large intestine, a disturbance in the distal part of the pulse is suggestive of disease of the anus, rectum or descending colon; of the middle part, of the transverse colon; and the proximal part of the pulse for the ascending colon. These qualities are difficult to appreciate.

3. A healthy person is one in whom the pulse flows smoothly, without turbulences or kinks, which has a certain tension but is yet

Normal pulse   Ropy pulse

FIG. 64

compressible and elastic, and which has the same characteristics throughout its depth (Fig. 64 left).

A person who has a pulse like that described above is, physiologically speaking, perfectly healthy: if he has had diseases in the past, they have become fully healed, nor has he any latent diseases which are due to become active and develop obvious symptoms, giving objective findings. This type of person is not likely to have any serious illness and will probably live a long time. If he becomes ill there will be a disturbance in only one or two places on the pulse; this disturbed pulse position, whatever other qualities it may have, retains its elasticity, signifying that the disease can relatively easily be cured. If a diseased pulse position has a certain hard and brittle quality, the disease is harder to cure.

It is taught that on occasions in the course of a serious disease and shortly before death, the pulses become normal.

Only on rare occasions have I myself felt a perfectly normal pulse in someone who was ill. This illustrates that although the pulse is accurate to an astonishingly high degree it is not, any more than anything else, a hundred per cent foolproof. For this reason, and as a double check, it is usually advisable to take a history, make a physical examination, laboratory investigations, etc. as suggested by the individual case.

A history, physical examination and laboratory investigations are useful in directing one's attention to what one might expect to find on pulse diagnosis. So much may be found on the pulse that it is useful to have some other means to act·as a pointer in discriminating between their relative importance. It should not be forgotten that the pulse divides diseases into twelve basic categories—and no more.

4. Diseases whether they be physical, physiological or mental show themselves on the pulse, provided the disease has a physiological effect.

If, for example, someone has had tuberculosis of the lungs, which has been fully healed so that the physiological function of the lungs is perfectly normal, the pulse of the lungs will be normal, despite the fact that a chest X-ray may show a few healed scars in the parenchyma of the lungs. The pulse is normal because the scars (unless extensive) do not influence the physiological function (and hence health and disease) of the lungs; just as the physiological function of the skin is not influenced (to any appreciable degree) by the healed scars of a few minor skin abrasions. If the tuberculosis were still active, or the healed scars so extensive as to influence pulmonary function, it would show on the pulse.

A diabetic will, if untreated, have an abnormal pulse. If his pulse is felt at such an interval after taking insulin that his blood sugar-insulin balance is perfect, the pulse will be so near to normal, that an abnormality will be missed unless one is specifically looking for it. If the pulse is felt a few hours before or after this ideal insulin balance, the abnormality of the pulse will be detected more easily, though of course not as easily as in the uncontrolled diabetic.

A substantial proportion of mental diseases, contrary to certain opinion, is really physiological, and hence can be treated by acupuncture. It is, for example, well known that anyone who is livery is liable to be depressed, in which case the depression can be cured by treating the liver. (The liver symptoms may not be present, but show themselves on the pulse of the liver.) A depression which has a purely circumstantial cause (e.g. bankruptcy) does not show itself on the pulse. Mental diseases are discussed in detail elsewhere.

5. A pulse which is ropy in outline (Fig. 64 right) and consistency is a sign of general, chronic, physiological unbalance. People with this type of pulse are very difficult to cure.

If a person who has enjoyed good health for most of his life becomes ill, the pulse of the diseased organ becomes abnormal, while the pulses as a whole, remain normal and smooth.

The person with a ropy pulse may not even have a specific disease that can be localised, though as a rule it occurs in people who, for many years, have taken drugs in excessive amounts, or whose habit of living has included anything else that might undermine their general health. Sometimes a ropy pulse occurs in elderly people who have had many illnesses affecting several bodily systems, all of which have been only partly cured.

6. The hollow pulse. Certain pulses are hollow, the examining finger feeling a normal resistance on light application, but immediately greater pressure is applied the finger, as it were, falls through into a hollow. This pulse signifies, in general, a deficiency state.

7. The wire pulse. Sometimes all the pulses, or only the superficial or deep, or even an isolated pulse become taut, hard and thin, like the E string of a violin.

This pulse signifies spasm and pain. In a patient who has pain or spasm, it is natural to tighten up, which also occurs in the pulse which becomes hard and tight like a wire.

The wire pulse is typically seen on the bladder pulse in lumbago or sciatica. Nervous people may have all their superficial pulses wiry.

8. A hard, round, incompressible pulse, usually signifies a stone, whether it be biliary or renal. Occasionally the size of a stone can be judged in this way, and if it is judged small enough to be able to pass along the biliary ducts, the gall bladder may be stimulated to expel it.

9. A blown up pulse may occur in the stomach due to aero-gastria; similarly the cardiac pulse may be blown up in cardiac strain or hypertrophy.

10. A coarse and rough pulse may be the result of cold weather. This condition disappears after the patient has been in a warm room for about half an hour.

11. The pulses become deeper in winter and in diseases associated with hardening and cold.

12. The pulses are more superficial in the summer and in febrile diseases.

13. Sometimes the pulse may be split longitudinally in two, which is more often noticed with the pulse of the gall bladder than with any other. I do not know the significance of this characteristic apart from the fact that it is associated with weakness. Possibly it indicates a non-synchronisation of the functions of the left and right biliary systems.

The Method of Utilising Physiological and Other Relationships

The more one feels the twelve basic pulses, the more does one have the impression that what is felt is not the specific organ concerned, such as the heart or the liver, but rather the basic 'conception' behind it, rather like Goethe's idea of the 'Urpflanze'.

Though obviously more than twelve organs or parts of the body can be effected by disease, with extremely few exceptions, they all show on the twelve basic pulses.

I think this can best be understood if the human being is considered as being in essence the result of the interplay of twelve basic forces which, in the course of embryonic development (and phylogenetic evolution), arrange the different individual cells and groups of cells to form twelve basic physiological and anatomical entities. The way that cells or groups of cells move during embryonic development along the most complicated paths is suggestive of some underlying force directing their movements.

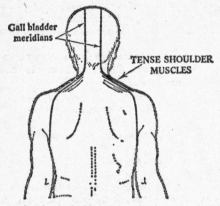

FIG. 65 Due to mental tension the shoulder muscles may tighten. The gall bladder meridian runs over these muscles; and hence the gall bladder or its coupled organ the liver, should be treated to relax the muscles

Embryologists explain this by the concepts of chemotaxis or polarity but these are probably only a very partial answer. Perhaps this can best be made clear by an example:

A disturbance in the liver pulse may be caused by:

(a) Disease of the liver itself—congestion, cirrhosis, carcinoma, etc.

(b) Haemorrhoids—the portal circulation, of which the haemorrhoidal veins are a part, passes through the liver. Hepatic congestion should always be treated first and only later the safety valve—the haemorrhoids.

(c) People who bruise easily. Presumably the clotting factors, such as prothrombim, are not sufficiently produced by a weak liver.

(d) Biliousness, nausea and vomiting.

(e) Migraine. Migraine (the usual type) is generally basically due to liver (and gall bladder) disturbance with the associated nausea. Most people with migrainous types of headache are first bilious and then some years later develop migraine.

(*f*) Weak eyesight, pain in or behind or around the eyes, black spots in front of the eyes, zig-zags, etc. This may be explained via the traditional relationship between the eye and the liver (see chapter on five elements).

(*g*) Excessive muscular tension, especially around the shoulders and neck (Fig. 65) (see chapter on five elements).

(*h*) Inability to wake up fresh in the morning, however early one has gone to bed.

(*i*) Certain types of asthma, hay fever, skin rashes and other allergic or 'stress' symptoms. Probably due to the manufacture of antibodies and other factors in the liver and spleen (Fig. 66).

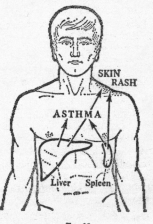

FIG. 66

The above-mentioned diseases may cause the disturbance of other pulses in addition to that of the liver. In addition most diseases have various causes, some of which will not affect the liver and therefore will not register on the liver pulse.

There are, of course, many more diseases which register as a disturbance on the pulse of the liver, but for the purposes of this discussion no more need be mentioned.

It will be noticed that most of the diseases or symptoms mentioned above (*a*) to (*i*) are related, (whether it be physiologically, anatomically, or by various laws of acupuncture) to the liver. The liver is the one factor that unites all these diseases, even though some of them may superficially not seem to have any relation to one another.

Similarly, if all the diseases of mankind are considered (excluding those few that have no physiological effect), it will be found that they all cause the alteration of one or several of the twelve basic pulses, and are hence related to the twelve basic organs. Anyone who is able to carry out the pulse diagnosis accurately, may prove the above statement to himself.

This fundamental conception that the human being is divisible into twelve basic physiological systems, to which other factors are subservient, is a gift for which we are indebted to the ancient Chinese of prehistoric times, for it is described in such detail in one of the oldest medical books in the world (The Nei Jing), that its origin must lie in a period that is even more remote.

When feeling the pulse, one has the impression, that what is felt is not the disease itself, but rather the entelechy behind the disease, showing itself only later as a disease in one or other branch of the basic twelve. This conception of the entelechy of disease is also suggested by the observation that the tendency to one of the basic twelve diseases can be foretold on the pulse diagnosis, months or even years before the disease or any symptoms or objective findings occur. This is the basic tendency which is only at a later date specialised into a specific disease.

FACTORS THAT SHOULD BE TAKEN INTO ACCOUNT WHEN PALPATING THE PULSE

Certain difficulties may be encountered in pulse diagnosis which may, in certain circumstances, necessitate relegating the pulse diagnosis to a secondary position.

Naturally the easiest pulse to palpate is that of the healthy middle-aged person, where the pulse is fairly elastic. In little children care must be taken because of the smallness of the pulse. In the aged plaques of atheroma in the radial artery may confuse the picture. It seems, as a rule, that if an atheromatous plaque is located specifically at a certain pulse position, then the organ system represented is diseased.

In certain people a dilation of the radial artery may occur at a particular pulse position. It might be considered that this dilation is merely secondary to a localised weakening of the radial artery and therefore dismissed. Experience usually shows, however, that the organ represented by the dilated segment is rather severely diseased and is fairly resistant to treatment.

In marked hypo or hypertension the pulse is either so uniformly weak or strong, that individual characteristics are difficult to palpate, unless the hypo or hypertension are first treated.

External influences must always be considered: Drugs, excessive food or drink, running, emotion, etc.

THE 28 MOST COMMONLY USED PULSE QUALITIES*

1. Floating
The pulse gives a sensation of floating on the surface of the skin; it responds to the finger when pressed lightly.

A Floating pulse is associated with Yang pulses, generally being found with the Exterior symptoms of Wind Evil lodging outside. If Floating but without strength, then it is a pointer to Empty symptoms.

2. Sunken (Deep)
The pulse gives the impression of being sunken between the muscle and bone; is only felt if pressed heavily.

A Deep pulse belongs to the category of Yin pulses, and is generally found with Interior symptoms of Evil Qi dormant internally. However, this pulse may also occur with Qi obstructed, Empty symptoms, etc.

3. Slow
The pulse is slow, only 3 beats to each breath.

The Slow pulse is also in the grouping of Yin pulses, generally to be found with inner organs Yin Cold symptoms. If a Floating pulse is associated with slow characteristics, then Yang is Empty outside; if a Deep pulse is associated with the Slow type, then Fire is decayed in the inside.

4. Rapid
The pulse is quick and feels urgent; 6 beats to each breath.

The Rapid pulse type is one of the Yang pulses; it is generally found with Fu Heat symptoms. If the pulse has little strength and

*The name in brackets e.g. (Deep) is an alternative translation as used in 'The Treatment of Diseases by Acupuncture'. In this book and whenever possible elsewhere I will not in future use the name in brackets, but only the main translation e.g. Sunken.

characteristics of the Floating and the Rapid type occur together, a disease of Yin Empty type is indicated; if on the other hand it is strong with at the same time the qualities found in Deep and Rapid types of pulse, then an over-abundance of Fire and Heat internally is denoted.

5. Slippery

The pulse is felt to come and go, with a flowing, round and slippery movement.

(a) If the blood is over-abundant, a Slippery pulse will be found. But pregnant women also have this type of pulse and in these cases it is not an indication of disease.

(b) The Slippery type of pulse can also be found in instances where the patient has Phlegm symptoms and where undigested food is left in the body tracts. It may also occur with Evil Qi extreme.

Whether this type of pulse does or does not indicate disease, can be established from other accompanying symptoms. Diseases associated with this pulse usually respond easily to treatment.

6. Choppy (Rough)

The movement of the pulse is felt as choppy and rough.

This type of pulse is found in cases where the Blood is deficient and the Jing injured. It may also occur in company with Qi obstructed or Cold and Damp symptoms.

7. Empty

The pulse comes with a floating and soft movement. It disappears on pressure.

This pulse is associated with Blood Empty symptoms. It may also be found in illnesses due to Summer-heat.

8. Full

The pulse feels full and forceful, long, big and hard.

This is a type found in cases of Evil abundant symptoms, and it may also occur with the

Fire symptoms of Evil abundant or Evil Full, obstructed and congealed.

9. Long

The pulse is felt as extraordinarily long, straight up and down, exceeding the extent of the original position.

This pulse is a sign of overflowing, and is present in symptoms denoting over-abundance, together with Qi rebellious Fire. If long, yet clear and round, as though feeling along a thin cane to the tip, it is a sign of health and not disease.

10. Short

The pulse feels short, choppy and little, seeming to have no head or tail, the middle arising suddenly and seeming unable to fill its natural place.

This is a sign of deficiency. It is also seen in cases where Original Qi is Empty and impoverished.

11. Overflowing

The pulse give the impression of being overflowing and large, filling the area below the doctor's finger; though abundant and large when coming, the movement loses its fullness and force as it passes.

The pulse is present in cases where symptoms suggest conditions of Evil Abundant and Fire Vigorous. If overflowing but without strength it suggests the cause of disease is Empty Overflowing and is a sign of Fire floating and Water dried up.

12. Minute

The pulse feels blurred, very fine and soft, almost imperceptible, as though about to be cut off, though it is not.

A pulse of this type when associated with Lost Yang symptoms, and where Qi and Blood are much decimated, cannot be reversed unless dealt with without delay.

13. Tight

The pulse is felt as taut and strong.

This pulse is found in conjunction with Cold

Evil symptoms. It may also be associated with pain symptoms.

14. Slowed-down

The movement of the pulse is broad, slow and even, with no great change in pace.

This pulse denotes the presence of stomach Qi, and in general is not a disease pulse. It may also occur in Damp Evil diseases.

15. Hollow

The pulse gives the sensation of being shaped like an onion stem; either floating or deep with a hollow area in the middle.

This pulse is present in cases where there has been great loss of blood such as haematemesis, epistaxis, menorrhagia, etc.

16. Bow-string (Wiry)

The sensation given by this type of pulse is that of pressing on the string of a fiddle, felt bouncing directly beneath the finger.

This type of pulse is found together with liver Wind symptoms. It can also be present with pain symptoms in Phlegm-fluid diseases.

17. Leather

This type of pulse is felt as large with a bow-string quality, yet hasty. It is obtained superficially but not when pressed hard. The sensation is that of pressing on the skin of a drum, the outside stiff and the inside hollow. Such a pulse denotes diseases due to an extreme excess of Exterior Cold. It is also found where there is loss of sperm or blood in men and miscarriage in women.

18. Firm (Hard)

This pulse is large, with a bow-string and hard quality. It can only be obtained deeply. This type of pulse is an indication of Qi-accumulation diseases.

19. Weak-floating

A pulse which feels very fine and soft, yet floating, only obtained when pressed lightly. Such a pulse is found with Yin Empty symptoms. It can also indicate the cause of disease as Kidney Empty, Marrow exhausted, where the Jing is impaired.

20. Weak

This pulse is felt as fine and small, yet deep. It can only be located when pressed heavily. When pressed lightly it is no longer felt.

This type of pulse occurs in association with Yang decayed symptoms. Where it occurs in the course of a protracted disease it should not be construed as a sign of danger.

21. Scattered

This pulse gives an effect of being floating, disordered and scattered; if the middle depth is selected, it gradually becomes hollow, and if pressed heavily, it ceases to be discernible.

This pulse is found in cases where the kidney Qi has been undermined and destroyed. There is always great danger in disease which shows a scattered pulse.

22. Fine

The pulse is felt like a thread, fine and soft, but more evident than a Minute pulse.

A pulse associated with decayed Qi. It may also occur where there are symptoms due to Damp. If found in conjunction with disease due to debility and absence of vitality, the disease is severe.

23. Hidden (Buried)

The pulse appears to be hidden. If you push the muscle and feel the bone then you begin to get its shape.

Such a pulse occurs in cases where the cause of disease has penetrated deeply into the organs. If the cause can be traced to a type of Yin Evil having been checked on the exterior and to the Yang Qi being isolated in the interior, then the pulse will become normal once sweating has taken place.

24. Moving

This pulse can be detected in the 2nd position, shaped like a bean. Its movement is slippery and rapid.

This pulse is associated with pain symptoms and is also present in disease due to fright.

25. **Hasty**   The movement of the pulse is sudden and hasty, halting at times.

This pulse is present in conjunction with Fire symptoms. It is also seen with disease due to an obstruction of Qi.

26. **Knotted**   The pulse moves slowly and gradually, sometimes halting.

This is a type of pulse which occurs in cases where the natural flow of the bodily process has been interfered with, causing an accumulation, congealing and obstruction.

27. **Intermittent**   The pulse is felt only intermittently. At times it almost stops, unable to regulate itself, and moves again only after a long time. Such a pulse denotes that the Zang Qi has been impoverished and destroyed. Its presence is indication that the disease has already become dangerous.

28. **Fast (Hurried)**   The pulse gives the effect of being hurried and anxious, 7 or 8 beats to each breath. It feels extremely agitated and urgent.

This pulse indicates a serious over-abundance of Yang Qi, and Yin Qi diminished to the point of exhaustion. It is a warning of the danger of sudden death.

Nowadays the 28 pulse qualities described above are often classified under 6 groups, as it is often nearly impossible to make some of the finer distinctions. There are also other methods of grouping the pulse qualities. The 6 are:

{Floating (1)*   {Slow (3)   {Slippery (5)
{Sunken (2)   {Rapid (4)   {Choppy (6)

*The numbers are the same as for the 28 pulse qualities overleaf.

FLOATING (1) ┬─ LEATHER (17) Floating and extremely forceful, like touching the skin of a drum.
            └─ WEAK-FLOATING (19) Floating and extremely weak, like silk in water.
             ─ FULL (8) Floating or sunken both forceful, bounces on palpating finger.
             ─ EMPTY (7) Floating or sunken both weak, recedes from palpating finger.
             ─ HOLLOW (15) Floating or sunken both big, the middle empty the surface full, like the stalk of a spring onion.
             ─ HIDDEN (23) Sunken and extremely forceful, the palpating finger must press as far as the bone to feel it.
SUNKEN (2) ──┬─ FIRM (18) Sunken and forceful, between floating and sunken in position.
            └─ WEAK (20) Sunken and extremely weak, the finger must feel for it carefully.

SLOW (3) ┬─ SLOWED DOWN (14) 4 pulse beats per breath.
         └─ KNOTTED (26) Slow and stops from time to time.
          ─ INTERMITTENT (27) Slow and rapid beats in an irregular pattern, after it has stopped followed by rapid period.
          ─ SCATTERED (21) The rate of the pulse is irregular, can only be felt in floating position.
          ─ MOVING (24) Rapid pulse in 2nd position, without head or tail.
RAPID (4) ─┬─ HASTY (25) Pulse rapid, stops from time to time.
          └─ FAST (28) 7 or 8 beats per breath.

SLIPPERY (5) ┬─ BOWSTRING (16) Like touching the string of a harp.
             ─ TIGHT (13) The pulse is tight like a twisted rope.
             ─ LONG (9) The pulse is neither big nor small, but long as a bamboo pole.
             └─ OVERFLOWING (11) The pulse feels big as it strikes the finger and weak long when leaving.
             ─ SHORT (10) Shaped like a bean, just touches the finger and then disappears.
CHOPPY (6) ──┬─ MINUTE (12) Extremely minute and soft, press it and it disappears.
            └─ FINE (22) Like minute but finer.

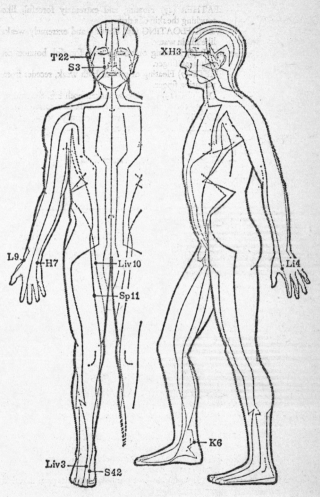

T22
S3
XH3
L9
H7
Liv 10
Sp11
Li4
K6
Liv3
S42

FIG. 67

According to the *Su Wen, sanbu jiuhou lun* pulse diagnosis may be performed on several of the superficial arteries. The body is divided into the upper (head), middle (hands), and lower (legs) parts; and each of these into the heaven, earth and man regions; the pulse in each case being felt at the named acupuncture point (Fig. 67).

	Heaven	Earth	Man
Upper body	XH3	S3	T22
Middle body	L9	Li4	H7
Lower body	Liv10 (women Liv3)	K6	Sp11 (stomach symptoms S42)

In dangerous diseases the prognosis can be arrived at by palpating the arterial pulse at the positions Liv3, K6, and S42. I have not met anyone who knows how to do the above.

X

THE CAUSE OF DISEASE

Since the 17th century, the general tendency in Western medicine has been to search for a purely physical cause in every human ailment. We attribute infections to bacteria or viruses. We consider that certain diseases (e.g. beri-beri, and some anaemias) result from deficiency states in the body and others from injuries to internal organs (as diabetes supervenes on an injured pancreas.) Even mental disease is either traced to a wholly physical cause, such as the secretion of an abnormal chemical within the body, or thought to be a condition of the mind produced by, and confined to, the mind alone.

In the Middle Ages, on the other hand, the cause of disease was more often seen as cosmological than physical, as directly due to forces outside rather than inside man. Sickness was the punishment meted out by Heaven to the evil-doer; the mentally deranged were possessed by the powers of darkness or suffering under a magic spell. The conjunction of stars and planets at a man's birth conditioned the physical weaknesses and the diseases likely to accompany them, which he would bear with him to the grave. The world in which man lived was malignant as well as benign; certain places were centres of healing but others were centres of disease; aches and agues emanated from subterranean streams and among the fruits of the earth grew also its poisonous plants.

In a sense, Chinese medicine combines these two attitudes. Among their many systems for the classification of disease, the most important is that which divides its causes into outer and inner 'influences', the former being meteorological conditions in the widest sense, the latter the emotions. Those relatively few diseases which fitted neither category were entered under miscellaneous headings.

The Meteorological Cause of Disease

When I was a doctor practising orthodox medicine, I used to think that changes in the weather had very little to do with disease; but, since becoming acquainted with Chinese theories on the subject, I have noticed how often, in a scarcely perceptible way, the weather actually does influence disease. It is, of course, a well-known fact that women with cystitis or an irritable bladder are more troubled in a cold season and that a particular type of migraine always occurs at mid-day in tropical countries. But the body seems to react to something more than simple changes in the climate. A patient suffering from that type of lumbago which becomes more acute in cold weather will notice that, though the temperature in his room may remain constant, his pain increases as the temperature outside falls. It would seem that a fall of temperature is not the only cause of his discomfort and that the body is influenced to just as great a degree by some subtle simultaneous change in the environment.

The Chinese divided the different meteorological conditions into what they called the 'six excesses', thus:—

WIND

Wind, since it belongs to the element 'wood', is the ruling element of spring (see chapter on five elements), and it is for this reason that wind diseases occur most often at this time of year. 'Evil wind' as the Chinese call it, is able to enter the body more easily when unaccustomed temperatures have dilated the pores of the skin.

The symptoms of 'wind injury' are coughing, headache, blocked or running nose, sneezing, etc. What is known as greater Yang Penetrating Wind brings dislike of cold, sweating, headache and a floating and slowed down pulse.

Wind may be associated with, or caused by, other 'evil' conditions. It may result from cold producing 'wind-cold' or originate from heat becoming 'wind-heat' and in both cases the consequent symptoms will be a mixture of the component elements. Such mixtures occur more often in wind conditions than in the other meteorological states. What is called 'inner wind' may also occur, producing 'phlegm-fire hot and abundant' or 'blood-empty wind-moving' symptoms of fainting, nervous spasms, vertigo, numbness, distortion of mouth and eyes with stiffness of the spine. Here the wind, since it is created within the body, does not belong to the wind of meteorology.

COLD

Cold, associated with the element 'water', is therefore the dominant of the six excesses during the winter, though it may also appear at other times of the year. It is a Yin evil and most likely to injure the Yang Qi of a patient.

When cold evil affects the exterior of the body, it may give rise to the symptoms of fever with a dislike of cold, breathlessness without sweating, headache, pains in the body and a tight and floating pulse. When the cold evil enters the meridians, it produces cramps and pain in the bones and muscles, and when it invades the solid and hollow organs, diarrhoea, vomiting, intestinal noises and abdominal pain.

But there is a type of 'inner cold' which, being created within the body, does not belong to the six excesses. This is known as 'Inner Zang Yang Qi empty and weak' and the symptoms (vomiting, diarrhoea, coldness of the limbs, rapid pulse and grey complexion) are caused by empty Yang creating inner cold.

'If the evil is in the stomach and spleen. . . the Yang Qi is deficient, Yin Qi excessive; the intestines therefore murmur and the abdomen is painful'.

(Ling Shu, wuxie pian)

SUMMER HEAT

Summer heat, belonging to the element 'fire', is naturally predominant at the full height of summer.

'In heaven there is heat, in earth there is fire. . . their nature is summer heat.'

(Su Wen, wuyunxiang dalun)

'On the extreme days of early summer there are warm diseases, on the extreme days of late summer, summer heat diseases.'

(Su Wen relun)

The chief symptoms of these summer heat diseases are headache, excessive body-heat, parched mouth, palpitations and consequent awareness of the heart action, spontaneous sweating and a rapid and overflowing pulse.

There are various types of this disease. A man who faints, after working hard or taking a long walk on a hot day, will be suffering from the type called 'penetrating summer heat', which comes under the category of 'Yang summer heat.' If, on the other hand, in the full heat of summer he has too cold a drink, the Yang Qi will be exhausted by the Yin cold and he will suffer from the 'penetrating cold of summer heat', in the 'Yin summer heat' category. This may give him an unpleasant sensation of chilliness, a dull ache in the head, abdominal pain with vomiting, and he may perspire freely. Summer heat may also combine with dampness to produce a red or white vaginal discharge, or dysentery, vomiting, abdominal pain and muscle spasm.

DAMP

Damp, belonging to the element 'earth', is the controlling meteorological condition of late summer. The various ailments which derive from it are classified as upper, lower, exterior or interior dampness, all characteristically heavy, muddy, greasy and blocking in their nature. They are often contracted after exposure to fog, after wading or working in water or being drenched with rain or from living in a damp place.

In 'upper damp' disease the head feels heavy, the nose is blocked, the face yellow and there is dyspnoea. In 'lower damp' there is oedema of the ankles and a thick vaginal discharge. 'Exterior damp' brings a sensation of alternating coldness and heat, spontaneous sweating, general weariness and lethargy; the joints ache and the limbs and body are swollen and puffy. 'Interior damp' is accompanied by a full, melancholic sensation in the chest, the vomiting of foul smelling food, swelling of the abdomen, jaundice and diarrhoea.

There is another type of dampness which is created internally, and hence not among the six excesses. It is caused by the consumption of wine, tea, cold melons or sweet greasy foods, all of which impede the function of the spleen.

DRYNESS

Dryness, belonging to the element 'metal', is the dominant meteorological state of autumn. A distinction is drawn between 'cool dryness' and 'warm dryness.'

'In mid-autumn it begins to grow cool, the west wind blows in an easterly direction and there is much wind dryness disease. . . If the weather is fair for a long time and there is no rain, the autumn Yang is sunny. Those who are affected by it will often have warm dryness disease.'

(Tongsu Shanghan lun)

The symptoms of cool dryness disease are a slight headache, a dislike of cold, coughing, absence of sweat and a blocked nose. In warm dryness disease the body feels hot, the mouth is parched, there is sweating, coughing, a pain in the throat and chest, blood in the sputum and sneezing with a dry nose.

In addition there is internal dryness, which is not among the six excesses. This is due to excessive loss of fluid from sweating, vomiting or diarrhoea, a condition in which the skin is dry, wrinkled and withered, the mouth and lips dry and cracked and the stomach fluid also dried up. What the Chinese call 'diabetes' ('wasting and thirsty disease' and others with similar symptoms), with hiccoughing, dry and hard stools, severe paralysis of the legs, convulsions and haemoptysis also occurs.

FIRE

The five meteorological conditions mentioned above can all, in extreme circumstances, be transformed so as to come under the element 'fire'. In most instances this will mean a more acute form of the disease. In, for instance, 'wind heat' disease the symptoms of a fixed look in the eyes, spasm of the limbs, curvature of the spine etc. are all due to the simultaneous increase in intensity of both wind and fire. The palpitations, facial flush, sweating, body heat and incessant thirst of 'penetrating summer heat' are really due to the transformation of summer heat to 'fire'. The exhaustion and depression which may accompany warmth and heat diseases will (if the change to 'fire' takes place) produce symptoms of burning lips and dry tongue, incoherent speech and mental confusion. If dryness changes to the element of fire, it will burn the lungs and cause coughing and blood-spitting. The deep red tongue and the awareness of heart-action, noticed in some fever patients, is caused by the cold evil changing to 'fire'.

In addition to, but distinct from, the above is 'internal fire', which has nothing to do with the six excesses. If 'wind' is very violent, liver fire is said to 'rise'. Many people in a fit of rage will go red in the face and feel a sensation of heat rising from the upper abdomen. Excessive eating and drinking will cause stomach fire to collect inside; excessive sexual activity causes the 'minister fire' to move wildly; deep feelings of grief or compassion will cause fire to arise in the lungs.

DORMANT QI

'If one is injured by cold in winter, then in spring one will have a warm disease. If one is injured by wind in spring, then in summer one will have diarrhoea. If one is injured by heat in summer, then in autumn one will have intermittent fevers. If one is injured by damp in autumn, then in winter one will have coughs.'

(Su Wen, yingyang yingxiang dalun)

The above quotation illustrates the Chinese belief that a disease, entering the body at one season, may remain dormant and not produce symptoms till a later season. Strictly then, a patient injured in spring (a wood season) by dormant wind (the excess pertaining to wood) will show symptoms of wind (wood) in the following summer.

A 'dormant evil' is able to enter the body because of a general weakness or lack of resistance in the patient; 'an empty place', as the Chinese put it, 'is that which contains evil'. For example, heavy manual work in winter will dilate the pores and cause sweating, so that 'cold evil' can easily enter through the skin. Similarly, a patient with weak kidneys is a ready prey to dormant 'cold evil' since the kidneys belong to the element water and are particularly susceptible to cold. The Chinese say that the dormant Qi is roused to activity by the body's response to seasonal changes: the Yang Qi begins to move outward in spring, the Yin Qi to move internally in autumn. Sometimes the dormant evil wakens of its own accord.

GENERAL CONSIDERATIONS

It goes without saying that everybody is not affected alike by the five meteorological conditions (the six excesses); indeed, some exceptionally healthy people are not affected by any of them.

Moreover, of two patients, the first may be attacked by one particular 'evil' while the second, immune from that 'evil', will succumb to another, just as one man will be afflicted by influenza, another by a duodenal ulcer, a third by cancer and a fourth remains perfectly healthy. As a rule the six excesses, like the dormant Qi, can only affect an already weakened part of the body. One of the aims of preventive acupuncture (see ch. XIII) is to keep the body as fit as possible so that these weaknesses are strengthened sufficiently early.

'*Wind, rain, cold and heat, unless they find the body weak, cannot by themselves do injury to man.*'

(Ling Shu, baibing shisheng pian)

Abnormal weather conditions are, from the meteorological point of view, the commonest cause of disease: sudden cold in summer, for instance, or warmth in winter. All doctors will have noticed that many patients are liable to fall ill when there is such a change in the weather, if it is particularly unusual for the time of year. If the low temperature is fairly consistent throughout the winter, flu or colds will be comparatively rare but a sudden fall of temperature at midsummer may well produce an epidemic of them.

'*Though wind Qi can create the myriad things, it can also harm them, as water can float a vessel but can also sink it.*'

(Yinkui yaolue)

Nor should the climate of the place in which one lives be overlooked. The cold and dryness of central Canada, the humidity of Singapore, the intermediate temperatures of England, where draughts and dampness are sometimes emphasised by inefficient heating, are all factors in the diseases from which different patients suffer.

The six excesses are able to invade the body only if the Protecting Qi is weakened. For the function of the Protecting Qi is to '*warm the flesh, moisten the skin, nourish the space between the skin and the flesh and control the opening and closing of the pores. It performs the function of protecting the exterior*'. The Nourishing Qi, which flows along the meridians, nerves and blood-vessels, is auxiliary to the Protecting Qi; if it is weakened it will in turn weaken the Protecting Qi, and here again disease is more likely to result.

The pulse, in pulse diagnosis, reflects the season and therefore also those adverse influences which might be expected at that time.

'On a spring day the pulse is floating, like a fish swimming on the waves; on a summer day it is superficial (in the skin), drifting like a surplus of the myriad things; on an autumn day it is below the skin, like an insect creeping into its winter shelter; on a winter day it is in the bone (deep), hidden like an insect hibernating.'

The Seven Emotions

In the previous section the outer (or meteorological) causes of disease were discussed. We now turn to the inner (emotional) causes.

Modern medicine might use the word 'psychosomatic' to describe the diseases considered in this section, as they are physical results of uncontrolled emotion; those in the previous section might be given the label 'somatopsychic', being mental diseases resulting from outer or physical causes.

As already mentioned, there is considerable interplay between the physical and the mental. An illness with a purely physical cause may produce both physical and mental symptoms; an illness of the mind may also produce disease in the body. Rheumatoid arthritis for, example, may have its insidious onset within the system, just like any other physical disease, or it may be precipitated by an emotional shock such as a broken engagement.

If the emotions are kept within the normal limits no disease results. If they are so powerful as to be uncontrollable, giving a man the feeling of being 'possessed' and of having his life governed by them, they will injure the body. A passing emotion, even a violent one, is harmless enough; if it is allowed to dominate the mind for any length of time it may well give rise to some physical disease. If a person, normally healthy in mind and body, is depressed by some distressing circumstance, he will recover from his depression as soon as the situation which caused it alters for the better. If, however, the situation persists for some time and the sufferer has some weakness which lowers his resistance, he will continue to feel depression even when its cause is removed. Because it has lasted for so long, the depression has either caused a weakness or aggravated some mild inherent one. Once this physical deterioration has begun it cannot be cured by purely mental means, such as will-power, except by a remarkably strong-minded person, and an easier course to adopt is first to correct the physical weakness. Then, provided the patient is capable of making some effort to help himself, the mental condition will right itself of its own accord.

The effect of the various emotions on Qi, the energy of life, are thus described:—

'*If there is anger, then Qi rises; if there is joy, then Qi slows down; if there is grief, then Qi dissolves; if there is fear, then Qi descends; if there is fright, then Qi is in disorder; if there is over-concentration, then Qi congeals.*'

(Su Wen, fengtong lun)

'*Joy injures the heart, anger injures the liver, over-concentration injures the spleen, anxiety injures the lungs, fear injures the kidneys.*'

(Su Wen, yinyang yingxiang dalun)

These quotations not only show the effect of the emotions on Qi but specify the particular organ injured by them, a point of practical importance. For, if anger injures the liver, then the patient who is always losing his temper over trifles can be cured by the correct treatment of his liver.

The heart is the central organ, as far as the seven emotions are concerned, and is often implicated by an emotional disturbance which primarily affects a different organ.

'*If there is grief and anxiety, the heart is affected; if the heart is affected, the five Zang and the six Fu tremble.*'

(Ling Shu, kouwen pian)

EXCESSIVE JOY INJURES THE HEART

'*When one is excessively joyful, the spirit scatters and is no longer stored.*'

(Ling Shu, benshen pian)

Most readers will, I am sure, have observed the fact that being deliriously happy or laughing uncontrollably does indeed give one the feeling described in the quotation as a 'scattering' of the spirit; one cannot, as it were, hold oneself together. On Chinese pulse diagnosis, people who laugh excessively will often (though not always) be found to have an overactive heart. In certain circumstances excessive joy can also affect the lungs.

'*If the lungs are joyful and happy without limit, then this will injure the animal soul.*'

(Ling Shu, benshen pian)

EXCESSIVE ANGER INJURES THE LIVER

If a naturally irritable person is faced with a problem he cannot solve or finds that his affairs have not gone according to plan, he may fly into a violent rage. Then '*Qi rebels and rushes upwards and anger and fire suddenly erupt*'.

> '*Qi and blood rebel upwards, causing man to feel joy and anger.*'
>
> (Su Wen, sishiji nicong lun)

The flushed, 'full-blooded' person, such as the conventional image of the butcher or the beer-swiller, is in fact more quickly roused to anger than the average man.

> '*If blood has a surplus then there is anger.*'
>
> (Su Wen, tiaojing lun)

Not only does anger come more easily to such a man than to the average but '*great anger can injure blood fluid and, if Yin blood is deficient, then water does not submerge wood, so that liver fire is even more vigorous.*' In these circumstances his body, which is deficient in Yin but full of 'fire', will react with fury to the most trivial provocation.

Although dysfunction of the liver is the typical predisposition, there are (as with everything else) often other causes of this abnormal anger:—

> '*Anger belongs basically to the liver. There are some though who say that anger originates in the gall bladder. This is correct if one considers that the liver and gall bladder are coupled organs, and although the liver is the stronger of the two, the decisions which are taken by the liver are essentially made in the gall bladder.*' (See 'The Meridians of Acupuncture', ch. XII)

> '*Some say that blood unites above, Qi unites below, and that one whose heart is troubled and alarmed is prone to anger. They consider that Yin is conquered by Yang, so that disease reaches the heart.*'

> '*There are some who say that, if the kidneys are abundant and angry for a long time, this will injure one's will-power. There are others who say that, if evil Qi invades the kidney meridian, the person in question is angry without reason, for anger starts in the Yin and then affects the kidneys. Hence anger may be produced by dysfunction of the liver, gall bladder, kidneys or heart.*'

> (Zangshi leijing)

EXCESSIVE ANXIETY INJURES THE LUNGS

If one observes oneself carefully, one will notice that a certain type of emotion, best described by the word 'anxiety', causes one to hold one's breath, to breathe shallowly, irregularly, or only in the upper part of the chest. This same type of anxiety can also affect the large intestine, since the large intestine and the lungs are coupled organs, and so produce ulcerative colitis, a condition to which over-anxious people are notoriously prone.

'In one who has anxiety the Qi (meaning both energy and air) is blocked and does not move.'

(Ling Shu, benshen pian)

Anxiety can also sometimes affect the spleen, for the lung and the spleen are respectively the arm and leg greater Yin organs and easily influence one another.

EXCESSIVE CONCENTRATION INJURES THE SPLEEN

The type of over-concentration referred to in this context is an obsessive concern with a particular problem. The sufferer from it cannot stop thinking about this problem; it occupies his mind from dawn till dusk and is still nagging at him as he falls asleep. He is almost literally buried as he broods on it behind locked doors, oblivious to the world around him.

A careful distinction must be made, however, between a man who is thus isolated by excessive concentration and one who is driven to this state by an inner fear.

EXCESSIVE GRIEF OR SORROW

Grief or sorrow may be caused by distress, vexation and suffering.

'If the heart Qi is empty, then there is sadness.'

(Su Wen, benshen pian)

'If Jing and Qi unite in the lungs, then there is sadness.'

(Su Wen, xuanming wuqi lun)

'If the liver sadness moves into the middle, then it injures the spiritual soul.'

(Ling Shu, benshen pian)

'If sadness is too extreme, then the pericardium is cut off from the rest of

*the body. If this happens, then Qi moves inside, and once this begins the
heart collapses.'*

(Su Wen, nue lun)

*'If there is sadness, then the pericardium is in distress; the lungs spread
and its lobes move; the upper warmer (lung and heart) does not penetrate;
nourishing and protecting Qi do not scatter; hot Qi is in the middle, so Qi
dissolves.'*

(Su Wen, futong lun)

These quotations show that the division of the seven emotions
into the five elements is not mathematically exact.

EXCESSIVE FEAR INJURES THE KIDNEYS

Fear can be aroused by a tension in one's emotional balance. Chinese
medicine held that one is particularly vulnerable to fear, if there is
some deficiency of kidney Qi or of Qi and blood or if the will-power
is weak and the spirit frightened. The kidneys are the storehouse of
the will-power, as is the heart of the spirit. If the blood is deficient,
then the will-power wanes; if the will is weak, one is prone to fear;
if one fears, then the spirit is frightened.

'If the spirit is injured, then fear is to blame.'

'If the Qi of the kidney meridian is deficient, then one is prone to fear.'

'If blood is deficient, then there is fear.'

EXCESSIVE FRIGHT

Fright is produced by a sudden encounter with the alarming and
unexpected, like the noise of a near explosion or the sight of an
appalling accident. This marks the distinction between 'fright' and
'fear', for fear does not happen suddenly; it is born from what one
already knows or expects.

*'If there is fright, then the heart has nothing on which it can rely, the
spirit has nothing to which it can turn, thoughts have nothing on which
they can settle. Hence Qi is in disorder.'*

(Su Wen, jutong lun)

Again, fright as one of the seven emotions does not fit into
the classification of the five elements. The above quotation shows
that it chiefly affects the heart; but, if the fright persists for any length

of time, it ceases to be either sudden or unexpected. It has become a fear, and will therefore affect the kidneys.

Miscellaneous Causes of Disease

EXCESS OF FOOD AND DRINK

The Chinese consider food to be the main source of energy in post-natal life. They believe that the stomach extracts the essence from ingested food and water and that this essence is then distributed to the rest of the body by the spleen, the coupled organ of the stomach.

Over-eating will affect the digestion and give a sense of fullness in the epigastrium, heartburn, a bad taste in the mouth, putrid breath, loss of appetite and irregular defecation. If food with too strong a taste is eaten, it may produce damp heat and phlegm, with the symptoms of belching, intestinal noises, excessive phlegm with a sensation of heaviness in the chest, and ulcers.

If one eats decaying food, one may get dysentery. If food is too cold, the Yang Qi of the stomach and intestines will be injured and the patient may suffer from stomach-ache, vomiting, swelling of the abdomen and a sense of suffocation and fullness in the epigastrium. Over-indulgence in hot and pungent foods may cause the stomach and intestines to accumulate heat and the stools to be dry and congealed. It may even result in bleeding haemorrhoids.

An excessive addiction to any one of the five flavours will suppress the function of the organ which it subjugates (see ch. VI). Too much sour (wood) food can injure the spleen (earth); too much bitter (fire) food can injure the lungs (metal); too much sweet (earth) food can injure the kidneys (water); too much hot (metal) food can injure the liver (wood); too much salty (water) food can injure the heart (fire) (Fig. 68).

The above system should be compared with that mentioned near the end of chapter VI where an excess injures its own element, i.e. 'sour injures the muscles (liver, wood)...' As with most things in Chinese medicine, sometimes one set of laws are followed and sometimes another. The experienced physician is the one who knows what will be the outcome with the particular patient he is treating.

Prolonged over-indulgence in alcohol can produce acute alcoholism or delirium tremens, and also a certain kind of general debility which renders the patient liable to many types of acute disease. Hunger and thirst sufficiently prolonged are also among the causes of disease.

PHYSICAL LABOUR

Man's body is so constructed that too little physical exercise will impair its health. In Chinese terms, the blood and energy flowing through the meridians and blood-vessels grows sluggish. On the

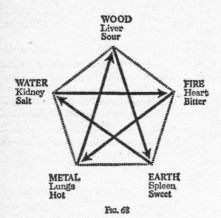

FIG. 68

other hand, excessive physical labour also has its dangers. It injures the spleen and can produce muscular exhaustion, lethargy in speech, breathlessness after exertion, slight fever, spontaneous sweating and abnormal awareness of the heart-beat.

'*If there is exhaustion, then Qi wastes away.*'

(Su Wen, jutong lun)

'*Looking for a long time injures blood; standing for a long time injures bones; walking for a long time injures muscles.*'

(Su Wen, benbing lun)

UNREGULATED SEXUAL ACTIVITY

Chinese philosophy, especially Taoism, which is so closely interwoven with Chinese traditional medicine, is largely based on the idea of the balance and harmony of life. Too much or too little in any direction will interfere with this balance. Thus too much or too little food, too much or too little exercise, and likewise too much or too little sexual activity, all these will endanger the harmony which is health.

> '*If sexual intercourse passes the norm . . . then it will injure the kidneys.*'
> (Ling Shu, xieqi zangfu bingxiang pian)

The 'essence' of life is stored in the kidneys. Excessive sexual activity weakens the essence of life and lowers the body's resistance to infection.

'*The kidney Yin and kidney Yang are deficient, or the Yin is empty and fire vigorous, or destiny fire is small and weak, or Yin and Yang are both injured.*' Such excess can produce the symptoms of coughing up blood, aching bones, hot flushes, night sweats, palpitations, lumbago, weak knees, coldness in the extremities, nocturnal emissions, impotence, irregular periods, menorrhagia or vaginal discharge. All the above symptoms have, of course, in many cases quite other causes; nor need any patient feel uneasy lest his doctor should take them as evidence of a life abandoned to lecherous orgies!

EPIDEMICS

The Chinese recognised the existence of epidemics and divided them into two main groups. The first group includes all those epidemics caused by abnormal climatic conditions: unseasonable cold or heat, excessive wind or rain, drought, floods or mountain mists at the wrong time of year. The second group consists of those epidemics due to poor hygiene: lack of cleanliness, dirt and refuse, putrefying bodies left too long unburied.

The evil energy (Qi) held responsible for these epidemics went under various names: Heterodox Qi, Perverse Qi, Confused Qi, Poison Qi, Demon Qi etc. These were thought of as invading the body from the outside, an idea which modern theories of bacteria or viruses as the causes of infectious disease do not contradict.

WOUNDS AND INSECT BITES

Under this heading come those diseases caused by accidents, snake-bite, insect stings and bites, rabies etc.

WORM INFESTATION

Chinese books usually give a fairly full description of diseases
caused by worms. Such diseases were common in China, owing to
the use of fresh, uncomposted night-soil. In my opinion Western
medicine is superior to traditional Chinese medicine in its classification and description of helminthic diseases, so it is unnecessary to
elaborate further on the topic here. One of the greatest benefits
Western knowledge has given to China has been the prevention of
helminthic diseases by adequate sanitation, drainage systems and
general cleanliness.

PENETRATING POISONS

Included in this group are all diseases due to the consumption of
poisonous berries, wood alcohol, bad or poisonous fish, bad meat
etc. In addition to the above are poisonous medicines, inaccurate
prescriptions, or overdoses of normally beneficial medicines.

HEREDITY

This class describes the ordinary hereditary diseases which are dealt
with in any Western book on medicine. It also includes shocks
suffered by the mother during pregnancy:—

*'If the man's mother had a severe fright while she was pregnant and the
Qi ascended but did not descend... then this would cause the child to have
epilepsy.'.*

(Su Wen, qibing lun)

XI

NEEDLE TECHNIQUE

The only thing of importance in acupuncture is to stimulate the right place. What the stimulus is, is of secondary importance.

Normally a needle is used, and this, in my experience, is the most effective. Massage, various types of electrical stimuli, mechanical vibrators, heating, magnetic oscillators have all been tried but are not quite as effective. In the Far East the pith of Artemesia Japonica (moxa) is dried and rolled into balls about two millimetres in diameter; one is placed on the acupuncture point of choice and lit so that it glows like the lighted end of a cigarette. This is an effective stimulus, but it may cause burns and even scars which do not necessarily disappear. This method, called moxibustion is supposed to be more effective in diseases due to cold and dampness, but in my experience this is not the case; and as it is no more effective than a needle I rarely use it. Another type of heating treatment, used in diseases due to cold and damp is to use the long handled type of Chinese needle. About an inch is cut off a moxa stick which is shaped like a cigar, and pushed over the exposed part of the needle. The moxa is lit and the heat is conducted down the shaft of the needle to the surrounding skin and flesh. As I find this no more effective than simple needling, I rarely use it. There are many old and modern variations to the above, but none are as simple and effective as a needle.

The needles may be made of any material. Silver alloys have the advantage of having some self-sterilising properties, which is an additional secondary safeguard. Stainless steel is best for thin needles as silver is too soft. Stainless steel needles have to be thrown away when they become blunt as they are difficult to resharpen. Silver needles can be resharpened on a very fine carborundum or other stone. The silver needles are best sharpened on several surfaces so that the tip is a cross between the cone of an ordinary sewing needle and the pyramid of a leather cutting needle. In this way they pierce the skin more easily yet do not cause bleeding as easily as a leather cutting or surgical needle. The much finer stainless steel needles should be sharpened like a cone, as is usual for ordinary needles. Injection needles may be used, but they easily cause bleeding and theoretically could harbour some dirt in the hollow of the needle; while a solid acupuncture needle, is, as it were, wiped clean on all its surfaces in its passage through the skin. If it is intended to leave the needle in place, it will be found that the head of an injection needle is rather heavy and pulls the needle out of place. I use a hot air steriliser. Small, cheap, automatic ones are sold in dental equipment shops.

Some European doctors differentiate between silver and gold needles founded on a misconception of tonification and sedation (see below). This may have arisen as a translating error as in Chinese the characters for gold and metal are the same. I have found no reference to it in the Chinese literature, though possibly it exists. Whilst in China, several doctors asked me what this new invention concerning silver and gold needles as used in Europe was all about!

Traditional Chinese works on acupuncture describe at great length about fifty different ways of inserting acupuncture needles, with names such as: 'burning mountain fire technique' or 'green dragon wagging tail technique'. These techniques involve the following: Inserting the needle 3 or 9 or 81 times; pointing the needle with or against the direction of flow of Qi along a meridian; twisting the needle clockwise or anticlockwise; inserting the needle fast and taking it out slowly as opposed to slowly in and fast out; inserting the needle in three stages and pulling it out in one as opposed to insertion in one stage and pulling out in three—and many more refinements. I have tried assiduously to find a difference between these methods, but have come to the conclusion that basically there is no difference except insofar as it includes what is said in the ensuing lines.

The size of the stimulus increases with:

1. A fat needle.
2. The deeper the insertion.
3. The more the needle is pushed up and down, so that the tip causes greater localised trauma.
4. A blunt needle or one with a hook on the end (both undesirable).
5. The more acupuncture points are used having a similar effect (sometimes has severe effect).
6. Leaving the needle in longer (doubtful).
7. Repeating the treatment at frequent intervals.

Many doctors think that the bigger the stimulus, the greater the effect; but just as often it is the very reverse. I have many patients who respond best to only one or two shallow pricks with thin, sharp needles, with the needle not left in place and the treatment repeated only infrequently. Certain constitutional types respond best to light treatment, others to heavy treatment, just as certain patients respond best to small and sometimes even microscopic doses of ordinary drugs while average doses of drugs may have no effect or make them feel ill. Because I recognise this great variation in individual sensitivity I have on occasions been able to successfully treat a patient by giving them a half to a tenth of the same medicine as their general practitioner was unsuccessfully giving them. Most chronic conditions I treat only fortnightly and finish the treatment at even longer intervals, for sometimes the effect of a treatment is only apparent after a week or more and if the second treatment is done before the effect of the first one is apparent, the two treatments may antagonise one another with either no result or a temporary worsening of the patient's condition. Acute conditions may be treated more frequently. Patients whom I see from abroad I of course treat at more frequent intervals; but it requires greater clinical experience and judgement on the doctor's part.

Chinese and European acupuncturists differentiate between tonification and sedation. Diagnostically one can say certain conditions represent underactivity whilst others represent overactivity. If for example the pulse is fine and weak one says it is underactive and requires tonification; if the pulse is strong and full one says it is overactive and the appropriate organ requires sedation. The Chinese and many Europeans also say that if the needle is inserted in a certain way, or one uses a silver needle, or one uses a point of sedation, that the appropriate organ is sedated; likewise if one inserts the needle in a different way, uses a gold needle, or a point of tonification, the same organ is tonified. I find on the contrary that whatever is done, as diagnosed on the pulse, the organ is brought nearer normality. If for example the pulse in the position of the heart is overactive (pulse full and strong) then whichever needle technique one uses, whatever the needle is made from and whichever point of the heart meridian one uses, the effect is the same: namely that the pulse becomes nearer that of a fine and weak pulse. Likewise if the pulse had been underactive (pulse fine and weak) and one had done exactly the same as above, the pulse would have become stronger. In other words the needle seems to exert a normalising influence: sedating the overactive, and tonifying the underactive; and if the doctor wishes it or not, he cannot (except under a few rare conditions) do the reverse. This normalising influence could fit in with the way the autonomic nervous system functions. It is interesting, at least philosophically, that overactivity and underactivity can be diagnosed, but that the treatment does not differentiate the two. Whether or not overactivity and underactivity are important from the point of view of Chinese traditional herbal medicine I do not know. In their theory it is important but perhaps not in reality.

The above jeopardises the whole idea of polarity, of Yin and Yang, coupled organs, hot and cold diseases, full and empty diseases, tonification and sedation, in fact much of the theoretical background to acupuncture. At the moment I prefer to sit on the fence for possibly both explanations have something to contribute. One might think it difficult to practise acupuncture without a firm theoretical background, but actually this makes little difference for when I treat a patient I do what I know works and nothing else. Likewise digitalis is used now in the same way as thirty years ago, even though the theory of its action has changed.

XII

DISEASES THAT MAY BE TREATED
BY ACUPUNCTURE—STATISTICS

Theoretically it is possible to help or cure by acupuncture any
disease that can be affected by a physiological process. Duodenal
ulcer, acne vulgaris, migraine, for example, are all the result of
physiological process, and as such may be cured: the duodenal
ulcer, by reducing the amount of acid produced by the stomach;
the acne, by increasing the function of the lungs and hormonal
regulation; the migraine, by increasing the function of the liver.

A trouble that is purely anatomical and uninfluenceable by a
physiological process, such as a kidney stone, advanced osteoarthritis, a fully formed cataract, cannot be treated by this means.
Human physiology is such that it is hardly ever possible for destructive changes in bones to be repaired—though it is obviously
possible to affect the circulation, swelling and muscle spasm around
an arthritic joint, without though altering the bone very much. In a
cataract the protein of the lens of the eye has become denatured,
a chemical change that cannot be reversed under the normal conditions of life.

The capacity for the regeneration of the new tissue in the human
being must be taken into account when judging the possibility
of a cure. It must be remembered that the human has less power
of regeneration than any animal, and vastly less than the lower
animals. A flat worm will completely regenerate itself if it is cut
in half longitudinally or transversely (so that two flat worms are
made out of one); if the tail of a rain worm is cut off, it will partially

regrow; the fin of a lung fish will grow again if it has been broken off; similarly the limbs of an amphibia (if cut under experimental conditions.) The human organism has not this same regenerative power. His creative energy has been transferred to the power of thought.

From the point of view of Chinese medicine, there is often no essential difference between a physical and a mental disease:

1. A physical dysfunction can cause a mental disease.
2. A mental dysfunction can cause a physical disease.
3. A physical dysfunction can cause a physical disease.
4. A mental dysfunction can cause a mental disease.

To give an example:

A If someone lives in depressing circumstances for a *short* time, then the depressing circumstances affect this mode of thinking, and he becomes depressed. If then after a short while the depressing circumstances disappear, then the patient can usually readjust himself mentally and the depression disappears.

B If, on the other hand, the depressing circumstances continue for a *long* time, then the depressed mind of the patient will eventually affect the liver (see relation between the liver and depression in the chapter on the five elements). Once the liver is affected, even if the original depressing circumstances are removed, the depression will remain. This depression can only be cured if the liver is treated—provided the depressing external circumstances have also been removed.

C If someone harms his liver, by, for example, eating a poison that destroys part of the liver, then the patient may (1) have predominantly physical symptoms such as jaundice, ascites, pruritis, or he may (2) have predominantly mental symptoms such as depression or an uncontrollable anger. In this instance whether the patient has physical or mental symptoms the liver would have to be treated physically.

To take another example of a mental disease which is really a physical disease and can therefore be treated by acupuncture: Most people who have 'various disturbances which include 'fear' have an underactivity of the kidney (many frightened children are bed wetters—kidney; after a fright most people wish to pass urine—

kidney). Actors with stage fright, teachers with lack of confidence, others before interviews, examinations, driving tests, people who are afraid to leave their house to meet strangers—all these are often kidney weaknesses which may be cured by treating the kidney. The law of the five elements, described earlier, indicates which organ is the culprit for the five main categories of mental diseases.

Case History. A regular colonel left the army at 45 and had to start civilian life. This involved interviews with many prospective employers. A certain degree of nervousness would be normal under the circumstances, but he was excessively nervous, so that he became unbearable at home. Pulse diagnosis showed an underactive kidney, which when treated reduced the excessive nervous tension, to one of normal proportions.

Case History. During the course of a takeover bid a patient's nerves were so shattered that he started shaking to such an extent that he could not pour out a cup of tea. The whole takeover bid was from his point of view jeopardised by his excessive nervous reaction, shaking hands, sleeplessness, fits of suicidal depression, etc. In his instance there was a dysfunction of the lung and heart which was quickly cured, and I am glad to say the patient's part of the firm prospered. He thanked me afterwards especially as he had been dragged to see me against his will by his wife.

One of the great contributions of Chinese medicine is its ability to link physical and mental diseases, whereby it is often found that a physical disease has a mental cause, and a mental disease a physical cause. In either case they may be treated by acupuncture.

In acupuncture psychologists would have a powerful weapon with which to treat their patients in a rational manner instead of rolling tranquillisers down their throats, passing electric currents through their brains, or discussing those parts of their sex life which they would rather forget. The role of the psychologist will still remain though in what has been classified above as 2 a mental dysfunction causing a physical disease, and 4 a mental dysfunction causing a mental disease. Quite apart from the above considerations, a good clinician should know which method (tranquillisers, E.C.T., psycho-analysis, acupuncture etc.) to apply in a given case.

The list of illnesses given below may be found in most books on acupuncture; some authors mention more, some less. Among the diseases listed some may be cured in nearly every case treated, while others may only yield to treatment in a small proportion of the patients treated. The duration of the disease, the amount of damage done, the general constitution of the individual patient must be taken into account. In many diseases, which have progressed too far for it to be possible to effect a cure, it is often possible to arrest the disease so that it does not progress any further; or a disease which is severely incapacitating may, at least, be partially cured or relieved, so that the man or woman may continue living a reasonably normal life.

Vague feelings of malaise, not feeling 100% fit but not really ill, not having enough energy or drive, etc., etc., are really all preclinical symptoms of disease, which, if they persist long enough, will quite likely result in actual disease. These vague preclinical symptoms can usually be precisely recognised by the pulse diagnosis (as discussed in the chapter on Preventive Medicine), and immediately treated. The increased sense of both physical and mental well-being that may thus be achieved, is a major contribution of acupuncture.

It is often said that 'the patient's psyche does not matter'. This is not true. It is not true for ordinary medicine nor for acupuncture, though some people (I think quite erroneously) consider that an objective diagnosis cannot be made unless the thoughts and feelings of both patient and doctor are disregarded, and a medical consultation is conducted like a test-tube experiment. It is well known to anaesthetists that the dose of anaesthetic required, especially for light anaesthesia, may have to be either doubled or halved according to the mental attitude of the patient—whether he wishes to become unconscious or whether he resists. The speed of recovery after an operation or the chances of life or death for a person who is very seriously ill are, as most doctors will agree from their own experience, partly a matter of the will power of the patient.

It is sometimes assumed that the mind is subjective, irresponsible and unreal, a negligible factor in medicine, since it cannot be measured; while the body is objective, measurable and real. To the acupuncturist they are but two facets of the same problem. Under certain circumstances one facet is more important, while at other times the other.

Some who have not experienced or seen the results of acupuncture get the impression that this is little more than hypnotism. This is by no means the case for:

(*a*) Acupuncture will work if the patient is completely unconscious under a general anaesthetic.

(*b*) There are certain sensitive people who notice the effect of a needle within a few seconds. As an acupuncture point is small it is occasionally possible to miss the exact spot, whereupon the sensitive patient, if he has already had experience of the treatment, will remark that it is not working. The acupuncturist can verify this by the pulse diagnosis. This should, in any case, be repeated as a routine after all the acupuncture needles are in place, for the pulse should alter within seconds of the needle being put in. If the needle is then readjusted by $\frac{1}{10}$th inch the sensitive patient will at once feel the difference, which may be verified by the pulse diagnosis.

(*c*) Occasional cases of spontaneous cures may be attributed to accidental injuries to acupuncture points. This is rarely the case as at any one time only a few acupuncture points are active—and they are small. An injury over a large area which may include one or several acupuncture points seems to have no specific effect. The stimulus must be localised to have an effect.

Tropical diseases, of which most European acupuncturists have no practical experience, are not mentioned in the index.

Various acute surgical emergencies such as appendicitis, peritonitis, etc. and various other potentially lethal diseases are not mentioned in the index, for although the acupuncturist may well be able to treat them (as is done in China today), most European acupuncturists will, as a matter of principle, not treat these diseases, as acupuncture is new to Europe. In addition, many of these acute emergencies can be well treated by orthodox medicine.

Some people say one cannot cure a chronic disease. This though is quite wrong. An expert acupuncturist may cure it as easily as: taking out a thorn, or wiping away snow flakes, or untying a knot, or pulling out a cork. Even if a disease is of long duration it can be cured; those who say it is incurable do not know acupuncture properly.

(Jia Yi Jing, Vol. II, Ch. 1a)

Duration of Treatment

The number of treatments required to effect a cure varies considerably. The average patient when seen by the acupuncturist for the first time does not, as a rule, have merely one disease, but in addition a variety of mild chronic complaints which do not incapacitate him, but simply make life less pleasant. It is a flare-up of these mildly chronic ailments that actually brings the sufferer to the doctor. The patient may have, for example, dyspepsia, a bitter taste in the mouth, frequent headaches, insomnia, brittle nails, and an irritable mood that he cannot control. One or more of these symptoms will have become acute; for instance, the dyspepsia may have developed into a duodenal ulcer.

All these symptoms (in one individual patient) including the duodenal ulcer, will take an average of about seven treatments to cure (or if a cure is not possible, to ameliorate).

Once a patient has been completely cured of his various ailments and the pulse is normal, provided he is seen by the doctor for a check-up every six months (as described in the chapter on preventive medicine), his basic health will then, as a rule, remain satisfactory. If he should then (with a basically sound health) develop an illness it can usually be cured in relatively few treatments—even a single one sometimes being sufficient.

Certain difficult diseases, especially if they have been in existence over a substantial portion of the patient's life, have a hereditary tendency or have resulted in anatomical changes, may easily take more than seven treatments to cure.

A disease of short duration, provided the causes, which may not be apparent, are also of short duration, will probably take less than seven treatments.

A very small proportion of patients who do not improve while they are being treated, may notice an amelioration or even a cure some months later—seemingly the healing process may be very slow.

Response to Treatment

The speed of response varies considerably from patient to patient, and with each disease.

Certain patients feel a response within a few seconds of the first needles being in place the first time they come for treatment. Others may have to be treated four or possibly even more times, for the first response to be felt.

The effect of a single treatment may be noticed during the treatment or several days later.

After a treatment nothing tangible may be noticed. At other times there may be an increase in energy, a lightness and buoyancy due to the stimulating effect of the treatment. In some people there is a great feeling of relaxation which may be followed by a pleasant drowsiness due to the sudden release of tension.

Occasionally, and in certain people, there is a reaction before the improvement starts. This may seem, if it is not understood, to be a worsening of the condition. A reaction may be compared to what happens in the case of an infection deep in the hand which first becomes an acute boil (seemingly a worsening of the condition— the reaction) before it discharges its accumulated pus to the exterior. The infection in the hand could also have been cured by the absorption of the infection into the blood stream. In the latter case the improvement would have progressed smoothly without a reaction. With or without a reaction the end result is the same, though the acupuncturist naturally always tries to effect a cure the smooth way. Certain chronic conditions however have, of necessity, to be brought to an acute stage (the reaction) for it to be possible to cure them. On rare occasions a reaction may be experienced after every treatment. The following letter illustrates an extreme example.

Dear Dr. Mann,   26.3.70
    The effects of your last treatment were so extraordinary I think you'd like to know about them!
    For 48 hours I suffered a great deal of severe pain—Codis twice in the night and once or twice next day failed to ease it and my spine and neck and back got worse and worse.
    At one moment on the second day I thought I was going to seize up as I've done before and I thought 'I *must* get to Dr. Mann' for a needling, only to be brought up short by remembering it was the needling that had brought it on.
    Then it passed and steadily I improved till every one of the symptoms—waves of nausea, livery-ishness, *pains went and even, for the first time in years, I found myself able to drink an ordinary amount of wine (strongly disapproved of for me by my medicos who prefer that I take 'purple hearts') without any subsequent heaviness or other liver symptoms.

*She was unable to walk far.

'Cheers!' Well done and thank you very much!
With all good wishes from a grateful and rejuvenated old lady.
Yours sincerely.

The improvement that is noticed during a course of acupuncture
does not follow a steady course. As a rule the degree of improvement and its duration increases with each treatment till the stage is
reached where the improvement persists and becomes a cure that
lasts. The improvement from the first treatment may last minutes,
hours or days, the effect lasting longer with each repetition. Some
patients improve rapidly at the beginning of treatment but may
take a long time to achieve that extra little bit that makes the cure;
others improve slowly at the beginning and then take a sudden
turn and are cured in no time. The majority follow an intermediate
course. Most often there are various ups and downs during treatment and there is rarely an absolutely steady improvement—
nature does not know straight lines. Not infrequently there is a
setback at some stage of the treatment, which is then overcome by
altering the acupuncture points used.

The final result rests with the individual doctor, his knowledge
and ability of the subject.

A list (admittedly incomplete) of diseases that may be cured or
helped is given below. In any individual case a certain disease
may better be treated by orthodox medicine or in combination
with orthodox medicine or by other means. The best course can
only be decided in dealing with an actual case.

HEAD

Neuralgia, headaches, migraine, fainting, trigeminal neuralgia (sometimes), tics, spasms, cerebral arteriosclerosis causing senility (only
early stage).

LIMBS AND MUSCULATURE

Fibrositis, muscular rheumatism, sciatica, lumbago, swelling,
discoloration, cramps, intermittent claudication, cold hands and feet
(sometimes), oedema, writers' cramp, weakness, some types of
trembling, neuralgia of shoulders and arms, tennis elbow, early
rheumatoid or osteoarthritis, weakness or feeling of excessive
heaviness of limbs, frozen shoulder.

DIGESTION

Duodenal and stomach ulcer, hyperacidity, gastritis, dyspepsia, inability to eat ordinary food, non digestion of food, no appetite, undigested stools, pale stools, eructations, wind, abdominal distension, bad breath, dry mouth, bad taste in mouth, heartburn, pyloric spasm, nausea, vomiting, rectal prolapse, constipation, diarrhoea, various types of colic, atony, perianal pain or itch, haematemesis, underfunction of liver, tender liver, hepatitis, chronic cholecystitis, colitis, ulcerative colitis, pancreatitis, nausea and vomiting of pregnancy, vomiting of children and infants, abdomen feels cold.

RESPIRATORY SYSTEM

Asthma, bronchitis, tracheitis, shortness of breath, pulmonary congestion, pulmonary oedema, recurrent colds, coughs and mild pulmonary infections.

CARDIO VASCULAR SYSTEM

Angina pectoris, pseudo angina pectoris, pain or heaviness over cardiac area, fainting, palpitations, tachycardia, bradycardia, arythmia, cardiac insufficiency, oedema, endo, myo-, and peri-carditis, certain valvular defects, high or low blood pressure, arterial spasm, phlebitis, haemorrhoids, lymphangeitis, adenitis, pallor, pins and needles, poor circulation, fainting, feels easily cold.

GENITO URINARY SYSTEM

Renal insufficiency, pyelitis, cystitis, some types of renal colic, lumbago, bladder irritation and spasm, bed wetting, lack of control of bladder, early prostatic hypertrophy.

SEXUAL SYSTEM

Pelvic pain, painful periods, irregular periods, flooding, vaginal discharge, vaginal pain, itching, menopausal trouble, hot flushes, ovarian pain, impotence, frigidity, sterility, lack of sexual desire, nymphomania, pollution, mastitis, menopausal loss of hair (sometimes).

EYES

Weak eyesight, tired after reading a book a short time, not optical

defects, black spots and zig-zags in front of eyes, pain behind or
around eye, conjunctivitis, blepharitis, iritis (sometimes), glaucoma
(sometimes).

EAR, NOSE AND THROAT

Hay fever, rhinitis, nose bleeding, sneezing, loss of smell (some types),
sinusitis, catarrh, gingivitis, tonsillitis, laryngitis, loss of voice, noises
in the ears (only early cases), pharyngitis.

SKIN

Acne, itching, eczema, urticaria, abscesses, herpes, neurodermatitis,
etc., may also help psoriasis.

NERVOUS SYSTEM AND PSYCHIATRIC FACTORS

Nervousness, depression, anxiety, fears, obsessions, timidity, stage
fright, neurasthenia, wish to die, agitation, outbursts of temper,
yawning, excessive loquacity, sleeplessness, nocturnal terror, many
neuralgias, facial palsy, neuralgia after shingles (sometimes), petit
mal (sometimes), trembling, trigeminal neuralgia (sometimes).

GENERAL STATE

Anaemia, general fatigue, lassitude, excessive perspiration, excessive
sleep, excessive yawning, sensitive to changes in temperature, travel
sickness, post operative weakness, weakness after severe diseases,
insomnia.

CHILDREN

Most of the more common diseases of children, excepting the infectious diseases. Children respond quickly. An important aim of acupuncture is to achieve and maintain healthy childhood, as much
ill health of later years can then be avoided. Bed-wetting, fear of
the dark, bad tempered or frightened states, inability to learn
properly at school, underdevelopment, stunted growth (sometimes),
cyclic vomiting and acidosis in infants.

GENERAL HEALTH

Most patients who have been treated by acupuncture notice a
considerable improvement in their general health. This is because
acupuncture can correct those minor disturbances in health which

are undetected by other methods of diagnosis, and which if they remained untreated could in later years easily turn into a serious overt and easily recognised disease. The sensitivity of Chinese pulse diagnosis (see section on Preventive Medicine) makes it possible to detect minor disturbances, enabling immediate treatment to be given at an early stage.

The above list of diseases might seem to some people rather long, as if acupuncture were to be regarded as a general panacea. It should be realised however that acupuncture is not a single drug, such as penicillin, which is therapeutically applicable to only a limited variety of infections. It is, on the contrary, a complete system of medicine which encompasses the whole field of therapeutics. In some books, a list of diseases that may be treated by acupuncture, three times as long as the above, is given. The list is only a representative selection, which mentions relatively more of the easier to cure diseases than the very difficult ones.

STATISTICS

All methods of treating disease will produce occasional good results. A method of real value will produce fairly consistent results with a reasonably high percentage of cures and improvements.

These statistics should be read with caution, for the interpretation of the results of treatment in medicine, and their incorporation into statistics is difficult. The liberal or the strict attitude of the analyst may make a not inconsiderable difference; nor should it be forgotten that most people do not have one disease, but a multiplicity of complaints. A certain doctor may have excellent results in diseases which according to these statistics have produced only failures; while in reverse this same doctor may have very poor results in diseases which these statistics suggest are easy to cure—doctors, as patients, have their strong and weak sides.

A The following statistics of Mauries* (Marseille) may act as a guide. It consists of all (625) the patients he treated in a specified period who had been previously diagnosed and treated by doctors other than acupuncturists, with little or no success. It does not include patients who visited Dr Mauries before they had seen another doctor; so that the possible statistical error of a spontaneous cure despite treatment is at least partly negated. Patients whom he has not been able to contact, or who have left his district, have not been included. Despite the fact that the diagnosis has been made by at least two doctors in each case, it will be obvious to the reader that some of the criteria used in diagnosis and the meaning attached to a particular diagnosis are a little different from those usually employed in England, for which due allowance should be made.

*(Actes des III eme Journées International d'Acupuncture).

Rheumatic and allied diseases	No. treated	Cured	Improved	Failure
Lumbago	29	19	6	4
Sciatica	25	15	4	6
Facial neuralgia	11	6	4	1
Rheumatism of several joints	36	19	11	6
P.C.E.	5	3	1	1
Cervical arthritis	9	5	3	1
Pain in heel of foot	2	2	—	—
Interscapular neuralgia	1	—	—	1
Gout of big toe	2	2	—	—
Tennis elbow	1	—	—	1
Arthrosis deformans	1	—	1	—
Generalised vertebral arthritis	3	2	1	—
Coccydynia	2	—	—	2
Arthritis of knee	14	9	1	4
Mandibular arthritis	2	1	—	1
Frozen shoulder	4	2	1	1
Rheumatism of knee and ankle	1	1	—	—
Coxarthritis	1	—	—	1
Intercostal neuralgia	3	2	—	1
Hernia of lumbar disc	1	—	—	1
Traumatic lumbar pain	1	—	—	1
Arthritis of shoulder	5	3	2	—
Cervico-brachial neuralgia	5	4	1	—
Post-menopausal rheumatism	1	1	—	—
Rheumatism of ankle	2	2	—	—
	167	98	36	33

i.e. 80% cured or improved

Pulmonary diseases	No. treated	Cured	Improved	Failure
Hay fever	9	6	—	3
Emphysema	10	3	4	3
Chronic bronchitis	3	1	2	—
Asthma	38	24	8	6
Cough due to hypertension	1	1	—	—
	61	35	14	12

i.e. 80% cured or improved

Urology	No. treated	Cured	Improved	Failure
Cystitis	2	2	—	—
Incontinence	8	2	3	3
Renal colic	1	1	—	—
	11	5	3	3

i.e. 72% cured or improved

_E.N.T._	No. treated	Cured	Improved	Failure
Streptomycin tinnitis	1	—	—	1
Chronic sinusitis	1	1	—	—
Chronic tracheitis	2	1	—	1
Chronic laryngitis	1	1	—	—
Chronic otorrhea	1	1	—	—
Catarrhal deafness	1	1	—	—
Allergic oedema of larynx	1	1	—	—
Post menopausal deafness	1	—	—	1
	9	6	0	3

i.e. 66% cured or improved

Gynaecology	No. treated	Cured	Improved	Failure
Dysmenorrhea	5	5	—	—
Hypermenorrhea	1	1	—	—
Menstrual trouble	1	1	—	—
	7	7	0	0

i.e. 100% cured or improved

Diseases of arteries and veins	No. treated	Cured	Improved	Failure
Arteritis of leg	3	—	2	1
Circulatory disturbance in a man	1	1	—	—
Circulatory disturbance in a woman	2	1	—	1
	6	2	2	2

i.e. 66% cured or improved

Cardiology	No. treated	Cured	Improved	Failure
Cardiac asthma	2	—	—	2
Paroxysmal tachycardia	1	1	—	—
	3	1	0	2

i.e. 33% cured

Digestive tract	No. treated	Cured	Improved	Failure
Gastralgia	3	3	—	—
Diarrhoea with food	1	1	—	—
Vomiting due to megaoesophagus	1	—	—	1
Peptic ulcer	1	1	—	—
Vomiting of infants	3	3	—	—
Chronic diarrhoea	2	2	—	—
Constipation	8	5	—	3
Biliary atony	6	5	1	—
Gastric ulcer	4	2	1	1
Habitual vomiting	1	1	—	—
Cholecystitis in a colonial	1	1	—	—
Gastralgia in a syphilitic	1	1	—	—
	32	25	2	5

i.e. 84% cured or improved

Neurology	No. treated	Cured	Improved	Failure
Littles disease	1	—	1	—
Results of hemiplegia	6	—	5	1
Myelitis	1	—	—	1
Para-facial spasm of Meige	1	—	1	—
Epilepsy	1	—	1	—
Tabetic pains	1	1	—	—
Parkinson's disease	3	—	—	3
Atrophy due to polio	1	—	1	—
Spasmodic quadriplegia due to cervical disease	1	—	1	—
Disseminated sclerosis	3	—	—	3
	19	1	9	9

i.e. 52% of improvements including one cure.

Endocrinal diseases	No. treated	Cured	Improved	Failure
Diabetes mellitus	3	—	—	3
Diabetes insipidus	1	—	1	—
Hyperthyroidism	1	—	1	—
Too short in stature	1	—	1	—
Adrenal insufficience	1	1	—	—
Pagets disease	1	—	1	—
	8	1	4	3

i.e. 62% cured or improved.

Neuro-vegetative disequilibrium (syncope, tachycardia, globus hysterias, spasms, lassitude, etc., etc.)

	No. treated	Cured	Improved	Failure
General neuro-vegetative disequilibrium	208	151	23	34
Post operative functional disturbances	3	3	—	—
Nervous hypertension	12	11	1	—
Neurasthenia	5	2	2	1
Vomiting of pregnancy	1	1	—	—
Angina pectoris	2	2	—	—
Neuritis of pregnancy	1	—	—	1
Pruritis ani	1	—	—	1
Eczema	1	1	—	—
'Floaters'	1	—	—	1
Demencia praecox	1	1	—	—
Plexalgia	1	1	—	—
Yawning	2	2	—	—
Excessive sleepiness	1	1	—	—
	240	176	26	38

i.e. 84% cured or improved.

Diverse diseases	No. treated	Cured	Improved	Failure
Quinche's oedema	3	3	—	—
Taenia	1	1	—	—
Excessive loss of weight	2	2	—	—
Amyotrophia	1	—	1	—
Furunculosis	2	2	—	—
Hiccups	1	—	1	—
Idiopathic headaches	6	3	2	1

Aphthous stomatitis	1	—	—	1
Non cardiac oedema of ankles	1	—	—	1
Seborrhea	1	—	—	1
Asthenia and anaemia	2	2	—	—
Bad at mathematics	4	4	—	—
Vertigo	1	1	—	—
Psoriasis	1	—	—	1
Obesity in a woman	1	1	—	—
Ophthalmic herpes zoster	1	1	—	—
Insomnia	3	2	1	—
	32	22	5	5

i.e. 84% cured or improved.

B The following statistics are taken from the Department of Surgery, Chung Shan Medical College, Canton, China.* They are concerned with the treatment of thirty-six cases of acute appendicitis, ten of appendicular abscess and three of perforated appendix with general peritonitis. They were treated mainly by acupuncture; though ten of them were treated by traditional Chinese herbs or a combination of both methods.

I. DURATION AFTER ONSET OF ILLNESS

Duration	Acute appendicitis	Appendicular abscess	Perforated appendix
2–6 hours	5		
6–12 hours	5		
12–24 hours	9	1	
24–48 hours	5		2
48–72 hours	4	2	1
4 days	2	1	
6 days	1		
7 days		2	
8 days		1	
10 days		1	
15 days		2	
Records unavailable	5		
	36	10	3

*Chinese Medical Journal 79: 72–76, July, 1959.

2. TEMPERATURE ON ADMISSION

Temperature	Acute appendicitis	Appendicular abscess	Perforated appendix
Normal	11		
High	24	9	3
37.1-38°C	17	3	
38.1-39°C	5	6	3
39.1-40°C	2		
Records unavailable	1	1	
	36	10	3

3. W.B.C. ON ADMISSION

W.B.C.	Acute appendicitis	Appendicular abscess	Perforated appendix
7,000 or less	4	—	—
7,000-10,000	8	1	—
10,000-20,000	19	6	2
20,000-30,000	1	2	1
30,000-40,000	1	—	—
Records unavailable	3	1	—
	36	10	3

4. SYMPTOMS AND SIGNS OF ACUTE APPENDICITIS CASES BEFORE TREATMENT

	No. of cases	%
Rigidity of abdominal muscles in right lower abdomen	24	66.6
Tenderness on pressure of right lower abdomen	36	100
Rebounding pain in right lower abdomen	36	100

5. CONDITION OF ACUTE APPENDICITIS CASES AFTER TREATMENT (Only those with complete record included)

Duration	Disappearance of abdominal pain	Normal blood picture	Normal temp.
<24 hours	9	7	12
<2 days	6	4	2
<3 days	5	4	3
<4 days	2	2	2
5-7 days	5	1	—
8 days	1	—	—

6. DURATION OF HOSPITALISATION IN ACUTE APPENDICITIS CASES

Hospital days	No. of cases		
Emergency cases not hospitalised	3		
2 days	4		
3 days	3	} 41.7%	
4 days	6		
5 days	8		} 78.7%
6 days	5		
8 days	1		
11 days	1		
12 days	1		
13 days	1		
22 days	1		
	36		

Conclusion: Good results obtained in all cases. No untoward complications were observed.

C L. J. Milman, E. D. Tikochinskaia and N. P. Bobrova (Acupuncture Laboratory of the Bechterev Psychoneurological Institute and the Polyclinic No. 5 in Leningrad)* treated thirty-five cases of physical sexual malfunction, which had proved resistant to ordinary treatment. Twenty-six of these were cured or improved.

*Russian Acupuncture Conference, Gorki, June 1960.

Of these twenty-six, twenty-four came for a re-check one-and-a-half years later; and of these twenty-four, twenty-one had remained cured or improved.

D Professor U. G. Vogralik (Gorki Medical Institute) states the following statistics:*

Disease	No. of Patients	Greatly improved or completely cured	Improved	No change	Treatment continues
Peptic ulceration	48	37	3	2	6
Spastic colitis	5	2	1	1	1
Bronchial asthma	54	3	31	14	6
Thyrotoxicosis (mild & severe)	12	3	6	1	2
Cardiac neurosis	16	0	5	8	3
Angina pectoris	18	7	7	4	—
Angina pectoris (sclerotic)	24	5	11	8	—
Rheumatic coronaritis	2	1	0	1	—
Erythraemia	23	6	11	4	2
Trigeminal neuralgia	13	4	4	1	4
Glaucoma	35	20	4	3	5
	250	88	83	47	

E At one time, when I practised acupuncture in hospital, we analysed and published the following results:†

The statistics given below refer to 40 consecutive patients seen at the Ear, Nose and Throat Department. In each case the main symptom was headache. We chose headache, as this symptom, with some exceptions, is difficult to cure by Western medical methods,

*Russian Acupuncture Conference, Gorki, 1959.
†Felix Mann and Anthony Halfhide. Medical World, April 1963.

while a reasonably permanent cure or considerable alleviation can be achieved by acupuncture in about 80 per cent of the patients treated.

Patients were referred to the Department by their general practitioners or from other departments of the Hospital. They were first seen by one of us (A.H.), and a thorough ENT investigation was made: this included almost invariably an x-ray examination of the sinuses. Where the headache was found to be due to ENT disease, such as sinusitis, and amenable to orthodox ENT methods, patients were treated by A.H. without using acupuncture. Such cases are not included in the figures. Other cases excluded were those in which the headache was of minor importance and cases of chronic suppurative otitis media, cranial tumour and so on.

All the patients had been treated for headache without much success by at least two doctors—their GP and a member of the ENT Department. Many of them had been to one or several other departments of the Hospital as well. Only those cases were treated by acupuncture (by F.M.) in which orthodox medicine had failed or achieved only a slight improvement.

We have not tried to classify the headaches into various types, as the usual definitions are too arbitrary and do not fit in with what we consider to be the important symptoms. Some doctors might have classified about half the patients as having migraine, other doctors, tension headaches, others neuralgia. Two of the headaches were specific—trigeminal neuralgia and supraorbital neuralgia.

The results, as recorded in the table, were as follows: 3 patients (7½ per cent) showed no improvement or aggravation; 5 (12½ per cent) showed moderate improvement; and 32 (80 per cent) were cured or showed considerable improvement. We have adopted this classification as many of the patients have had headache for a large part of their lives and the cause usually goes back several years further. Those very severe cases—those patients who before treatment had spoken of suicide, had stopped work, or were living in a semi-conscious state under constant analgesia usually still have an occasional mild headache, and so could be described perhaps as 70 to 99 per cent cured. A few with moderately severe headache still have a very occasional mild one—much as a patient cured of pleurisy may have an occasional pleuritic pain in cold or damp weather.

Table Analysis of cases treated by acupuncture for various types of headache and other symptoms or diseases

		Various types of headache				Other symptoms or diseases in the same patient			
Patient	Sex	Duration of headache (years)	Number of treatments	Result		Symptom or Disease	Duration (years)	Number of treatments	Result
J.D.	F	1	1	+ +		Vertigo	1 4/12	1	+
M.E.	F	7	1	+ +					
N.G.	M	5	2	+ +					
M.H.	M	10	12	+		Asthma	14	15	+
R.L.	M	15	7	+ +		Lumbago	?	?	+ +
D.L.	F	2	14	+ +		Biliousness	2	14	+ +
L.O.	F	2	6	+ +		Heartburn	2	6	+ +
B.P.	F	1½	5	+ +		Tinnitus	8/52	3	+ +
M.Q.	F	½	2	+ +		Asthenia	2	9	+ +
H.S.	M	8	4	+ +					
E.S.	F	8	4	+ +					
E.T.	F	20	8	+ +		Allergic rhinitis	20	8	+
P.T.	M	4	9	—					
H.T.*	F	3	1						
A.W.	F	20	4	+ +		Anosmia	2	4	+
H.N.	M	1	14+	+ +		General debility	1	14+	+
J.T.	M	10	7	+ +		Allergic rhinitis	10	7	+
D.H.	M	20	4	+ +		Biliousness	?	7	+ +
D.B.	M	2	10	+ +		Asthenia	?	10	+
P.C.*	M	1	1						
R.D.	M	15	3	+ +					
M.G.	F	1	8	+ +					
F.E.	M	10	8	+ +		Hypochondria			
G.C.	F	?1	12	+ +		This was a case of trigeminal neuralgia			
A.H.	F	9/12	3	+ +		This was a case of supraorbital neuralgia			
G.H.	F	6	3	+ +					
R.H.	F	3	6	+ +		Hysterical aphonia	30	6	+ +
M.K.	F	3	15	+		Allergic rhinitis	3	15	
T.L.	M	9/12	4	+ +		Asthenia			
K.M.	M	5	2	+ +					
J.M.	M	4	4	+ +		Allergic rhinitis	?	4	+
A.N.	F	12	8	+ +		Biliousness	12	8	+ +
E.S.	F	3	7	—					
M.S.	F	20	12	+ +		Allergic rhinitis	20	12	+
A.W.	F	5	3	+ +		Biliousness	5	3	+ +
L.P.	F	5	2	—					
E.A.	M	17	6	+ +		Biliousness	?	6	+ +
C.L.	F	8	9	+ +		Anosmia	?	8	+
D.M.	F	1	4	+ +		Very sleepy	1	4	+ +
A.M.	M	15	17	+					
V.H.	F	5	11	+ +					
R.W.	M	10	8	+ +					

+ + cured or considerable improvement
+ moderate improvement
— no improvement or aggravation
* did not continue treatment

Some patients, particularly those with severe symptoms of long duration, will probably have a mild recurrence of their headaches after several months or several years of freedom. These can usually be cured in one to three treatments. In the most difficult case there may be several recurrences, each being milder and separated by a longer interval until, in the end, the condition is completely cured, or at least nearly so.

In certain patients with headache due to diseases which cannot be cured by either Western medicine or acupuncture (there are none in this statistical series), it is sometimes possible to alleviate the headache. As the basic condition cannot be cured, constant 'pep up' treatments will be required *ad infinitum*—helpful perhaps but unsatisfactory.

As can be seen from the table, many of the patients had other symptoms or diseases which were treated at the same time as the headache. (Only the main additional symptom is noted in the statistics.) Mostly the treatment of the other symptoms took the same number of treatments as the headache—thus patient L.O. (7th down) needed a total of 6 treatments to cure both the headache and the heartburn. Sometimes the other symptoms took longer or shorter to cure or alleviate than the headache. Where there was little difference in the number of treatments required to treat both symptoms, the greater number of treatments has been put down under both headings. Where there is a question mark the relevant fact had not been noted in the case history.

F Ten of us* analysed the results of treatment of 1000 consecutive patients. The more rarely treated conditions were not included, so as to obviate spurious conclusions from a small review. Sometimes it was difficult to know under which heading to put a single patient for several had a multiplicity of diseases and symptoms. As a rule the disease for which the patient primarily went to the doctor has been included in these statistics.

*Felix Mann, Raymond Whitaker, Bernard Perlow, Robert Graham, James Rentoul, Gerald King, Rankin Martin, Henry Kobner, Lawrence Hyman, David Blake.

ANALYSIS OF 1,000 CONSECUTIVE PATIENTS TREATED BY
ACUPUNCTURE COMPILED BY 10 DOCTORS

	Total number of patients	Cure or considerable improvement.	Moderate improvement.	No result or slight improvement.
Headache, migraine	119	71	22	26
Neuralgia, trigeminal neuralgia	12	2	8	2
Lumbago, sciatica 'slipped disc'	67	31	16	20
Organic neurological diseases	30	6	5	19
Gastritis, peptic ulcer	41	15	18	8
Constipation	9	4	2	3
Colitis	11	6	2	3
Haemorrhoids	9	6	2	1
Osteo-arthritis, rheumatoid arthritis, ankylosing spondylitis	196	52	89	55
Muscular rheumatism, brachial neuralgia, cervical spondylosis	95	33	4	8
Gout	9	5	3	1
Metatarsalgia, tennis elbow	8	7	0	1
Asthma	57	25	17	15
Chronic bronchitis, pulmonary dyspnoea	11	6	3	2
Hay fever, vasomotor rhinitis	51	20	20	11
Tonsillitis	10	5	3	2
Angina pectoris	8	2	3	3
Intermittent claudication	5	0	3	2
Cramp in calves	7	5	0	2
Hypertension	10	0	5	5
Varicose veins	5	0	3	2

	Total number of patients	Cure or considerable improvement.	Moderate improvement.	No result or slight improvement.
Other organic vascular disorders	6	0	1	5
Cystitis, irritable bladder	8	5	1	2
Prostatism	9	3	3	3
Dysmennorrhoea	26	14	4	8
Vaginal discharge	11	7	1	3
Hot flushes	9	5	2	2
Nausea and vomiting of pregnancy	7	5	2	0
Impotence	27	7	7	13
General psychiatric	30	9	11	10
Claustrophobia	6	4	1	1
Unfounded fears	9	5	2	2
Endogenous depression	13	6	4	3
Excessive tension	16	8	4	4
Mental apathy	11	5	3	3
Insomnia	11	4	3	4
Premature senility (early stage)	15	10	1	4
Liver dysfunction	11	6	3	2
Food allergy	8	4	2	2
Car sickness	9	7	0	2
Acne	10	5	3	2
Blepharitis	6	4	3	0
Urticaria	5	4	1	0
Tinnitus	7	1	1	5
	1000	439	290	271

G The following 1518 cases were treated by Dr Johannes Bischko, in the E.N.T. department of the Vienna Allgemeine Poliklinik which is under the direction of Prof E. H. Majer. The period of observation was from 22.12.58 to 21.4.71. Of the 1518 patients treated, 1037 had a single disease whilst 481 had two diseases. All patients were treated in the out-patients department. Those patients who did not continue treatment have been added to the no result column.

Diagnosis	No. of cases	No. of treatments	Good result	+ —	No result
Tinnitus	309	16 to 25	105	124	80
Symptoms after middle ear operations	74	10 to 15	21	7	46
Lesions of inner ear	36	10 to 20	4	9	23
Presbycusis	72	8 to 16	27	32	13
Vertigo	96	8 to 15	66	19	11
Menière's disease	36	10 to 15	18	12	6
Facial paralysis	57	10 to 25	23	23	11
Anosmia	8	10 to 20	6	1	1
Hayfever	25	6 to 15	17	5	3
Sinusitis	259	4 to 11	200	36	23
Symptoms after sinus operation	84	5 to 10	33	19	32
Trigeminal neuralgia	118	2 to 20	73	31	14
Vasomotor headaches	189	4 to 10	129	43	17
Migraine – cervical	117	4 to 13	81	19	17
Migraine – ophthalmic	13	4 to 15	7	4	2
Migraine – vascular	88	4 to 20	63	15	10
Spastic torticollis	17	10 to 30	14	3	—
Stutter	71	4 to 15	48	13	10
Spastic dysphonia	102	4 to 15	66	32	4
Bronchial asthma	38	6 to 15	25	7	6
Ozaena	9	4 to 12	4	3	2

XIII

PREVENTIVE MEDICINE

In ancient China a first class physician was one who could not only cure disease but could also prevent disease. Only a second class physician had to wait until his patients became ill so that he could then treat them when there were obvious symptoms and signs.

It is for this reason that the doctor was paid by the patient when he was healthy and the payment was stopped when he was ill. This was so much so that the doctor had to give the patient free of cost the medicines required, medicines which he, the doctor, had paid for out of his own pocket.

'To administer medicines to diseases which have already developed and to suppress revolts which have already developed is comparable to the behaviour of those persons who begin to dig a well after they have become thirsty, and of those who begin to make their weapons after they have already engaged in battle. Would these actions not be too late?'

(Su Wen, Ch. 2)

This type of preventive medicine is based in acupuncture on the pulse diagnosis, which, as already mentioned, presents in its early stages, rather the entelechy of disease than the disease itself.

It is well known that the person who will at a later date develop, for instance, hypertension, exhibits certain mental symptoms (a certain stiff walk and fixedness of ideas), many years before the hypertension as such shows itself and similarly with other diseases. This type of preclinical symptomatology is so vague and uncertain that, on the whole, little use can be made of it.

The pulse diagnosis, on the other hand, is a certain indication of preclinical disease. A little consideration will show that before a disease develops with physical signs and objective findings, there will be, possibly for months or years beforehand, some physiological disturbance that is too slight to cause overt symptoms. But, even at this stage, the pulse registers a definite abnormality.

The acupuncturist will at this preclinical stage treat the patient, using acupuncture points dictated to him solely by the pulse diagnosis.

For preventive medicine of this type to be effective, the patient must, of course, see the doctor at regular intervals. In my experience a healthy person with a reasonable constitution need only have his pulse felt every six months; which is the same interval at which one should visit a dentist. Once a year is enough for the exceptionally healthy.

One additional advantage of this preventive routine is that the whole general level of health is maintained at a higher level. It can give not only the absence of disease, but a positive feeling of well-being with an abundance of physical and mental energy.

Very often people are not actually ill, but feel a little below what they sense should be ideal health. This is, in reality, the preclinical stage of disease which may take very many years till it is seen as an overt disease. If this person is correctly treated, not only is the slight fatigue, etc. cured, but in addition he is spared the consequences of later developing an obvious disease.

The length of life may be increased with less cerebral sclerosis and its attendant evils:

The Yellow Emperor once addressed T'ien Shih, the divinely inspired teacher: '*I have heard that in ancient times the people lived to be over a hundred years, and yet they remained active and did not become decrepit in their activities. But nowadays people only reach half of that age and yet become decrepit and failing. Is it that mankind is degenerating through the ages and loses his original vigour?*'

Qi Bo, the chief physician, answered: '*In ancient times those people who understood the ways of nature, patterned themselves upon the Yin and the Yang...*'

(Su Wen, Ch. 1)

When treating patients with early signs of ageing and cerebral sclerosis (such as apathy, inability to follow an argument and feeling generally decrepit), the response (if the process is not too advanced) is often remarkable, with a return of mental agility and youthfulness which is noticed by friends who do not even know that the patient has seen a doctor. Nevertheless, earlier diagnosis and treatment, as in preventive medicine, would probably have given even better results.

Parents who are healthy normally beget children who are also healthy. This is one of the most important aspects of preventive medicine for children who are born robust and healthy, have, as a rule, less disease later in life and on the whole a healthier mental outlook. If adults become ill, who were robust as children, they are easier to cure. The diseases which are hardest, and sometimes well nigh impossible to cure completely, are in those who were born weaklings and who were ill during the first few years of life.

For preventive acupuncture to be effective, the initial treatment must have been successful, so that the pulse has become normal. It may not be possible to completely cure someone who has been ill for many years; though the patient may say he is cured as he has no symptoms. In this case the pulse will still show a slight abnormality which cannot be corrected. In this type preventive acupuncture is only partially effective. If the patient had been seen earlier, preventive acupuncture would have been fully effective.

We have to face the fact that, in our modern civilization, with its many influences which are detrimental to health, by no means all diseases can be prevented though most acupuncturists find that not only is the general level of health heightened, the basic germ plasma of the next generation improved, and expectation of life increased, but also a substantial proportion of disease prevented. This type of preventive medicine applies particularly to the chronic and degenerative diseases and not to such an extent to the infective diseases or those caused by external agents, except insofar as the general resistance has been increased.

Various additional factors should not be forgotten: Exercise, naturally grown food that is not poisoned or devitalised, good air, enough relaxation and enough thought:

Modern man drinks wine like water, leads an irregular life, engaging in sexual intercourse while he is drunk, thus exhausting his vital forces:

they do not know how to preserve their vital forces, wasting their energy excessively, seeking only physical pleasure, all of which is against the rules of nature. For these reasons they reach only one half of the hundred years and then they degenerate.

(Su Wen, Ch. 1)

XIV

SCIENTIFIC VERSUS TRADITIONAL
ACUPUNCTURE
SOME CONCLUSIONS

The reader of chapters I and XI will see that in several essentials I
have distanced myself from the traditional conception of acupuncture points and meridians. Further, that the many categories of
acupuncture points described in chapter VIII are somewhat superfluous; and that the laws of acupuncture mentioned in chapters V, VI
and VII are so all embracing that they actually elucidate rather little.

Qi, the energy of life, and similar conceptions mentioned in
chapter IV, are more easily understood from a traditional point
of view.

This limitation of classical Chinese theory is bewildering to many
readers and indeed to numerous doctors who practise acupuncture.
Hence I thought it appropriate to mention this in greater detail in
the 1973 reprint. I felt the conclusions to be drawn from chapters I
and XI were obvious and thus had previously only mentioned them
in a few words.

In my neurophysiological theory I have explained the areas used
for stimulation in acupuncture. They are partially on a roughly
dermatome basis; partially involve "long" reflexes to distant parts
of the body, which implicates a distribution by specific spinal segments or nerves; and are partially via unknown connections.

This theory would transform the classical small specific acupuncture point into an area as large as that of a dermatome, or to
the distribution of a specific nerve, or even to an area of several
dermatomes if the area has previously been hypersensitised (see

pages 24 and 25). If only a few neurones are involved, the skin area could be considerably smaller than a dermatome.

In most instances no doctor, even an expert in acupuncture, can find an acupuncture point in those areas where there is a big expanse, such as the abdomen, back and thorax. If a group of doctors is asked to locate a specific acupuncture point in such an area, their positions will quite often vary by a considerable amount, and yet all these doctors are able to help or cure a large proportion of their patients, provided they have a disease amenable to acupuncture. This suggests to me that small specific acupuncture points rarely exist, and that those researchers who have found specific types of specialised nerve endings or other structures at acupuncture points are mistaken. The structures found by these histological investigations may well be there, but they do not correspond to acupuncture points, for they do not exist. Stimulation of any layer can be effective, whether it be skin, subcutaneous tissue, muscle or periosteum. Hence one should not speak of a dermatome, but rather of a dermo-myo-sclerotome. This poses some problems, for the different layers do not always have the same segmental innervation.

In a disease of the viscera or other parts of the body there is often a reflex tenderness in the associated part of the surface. This tenderness may include muscle spasm or circulatory changes. It also presumably affects most histological structures throughout the entire depth of the appropriate area, due to their similar innervation.

As far as I know, there are no specific histological elements in McBurney's point, which becomes tender in appendicitis. I think nearly every single part of the body can become reflexly tender, in a way similar to that of McBurney's point. Hence the number of acupuncture points becomes infinite—indeed some books mention so many acupuncture points that one wonders if there is any normal skin left.

McBurney's point is not a small discrete 'point,' but quite a large area, whose position is somewhat variable. McBurney's point lies in the appropriate dermatome. The remainder of the dermatome is not tender, or only mildly so, for as Kellgren (Fig. 5) and others have shown, certain areas within a dermatome show greater changes than others.

Some acupuncture points seem to have a constant position and may be tender even in a completely healthy person:

G21 is situated where the trapezius arches over the first rib and hence is presumably under greater tension than other parts of the muscle.

Sp9 is located below the medial condyle, over the lower part of the medial ligament of the knee, where many women have a tender oedematous area. As this usually occurs only in women, apart from those who have injured their knee, it is presumably hormonal. In some women this area becomes an oedematous pad of fat the size of a hand.

G20 is next to the greater occipital nerve where it arches over the occiput, just as B2 is adjacent to the supratrochlear nerve where it passes over the supraorbital ridge.

All the above and a certain number of other acupuncture points are nearly always tender, even in the healthy subject. This is probably often due to compressing a nerve trunk against the bone. Other places may be tender due to muscular tension sensitising the area and thus requiring a smaller stimulus from the acupuncture needle to be effective.

H7 is a more effective point than H3, as stimulation of H7 involves the needle piercing thicker skin and a hard ligament. This causes greater pain than needling the fatty tissue around H3 and thus obviously has stimulated more nerve fibres. For a similar reason acupuncture points which involve stimulation of the periosteum have usually a greater effect than those involving only subcutaneous fat, unless the needle is strongly twisted in the skin.

Stimulating a nerve trunk, which produces a lightning pain, is by no means more effective. In patients who have the so-called cervical disc syndrome and allied conditions, stimulation of the transverse process of the 6th cervical vertebra is more effective than trying to needle the adjacent nerves of the brachial plexus.

In my experience, contrary to classical theory, the type of stimulus used in acupuncture is of little importance, whether it be a needle, a thorn, an electric current, heat, a vibrator or injections. This would agree with the "all or nothing" response of nerve fibres, which either respond or do not respond to stimulation, there being no qualitative difference. The stronger the stimulus, the greater the effect, due to activation of a larger number of neurones or their repetitive stimulation. The traditional theory that there is a qualitative difference between a hot or a cold needle, or the manner in which it is twisted or inserted, does not concur with my experience and would be harder to explain neurologically. Sometimes if the periosteum is stimulated in the region of a joint the effect is greater than if the overlying skin is needled. Possibly this is due to activation of a local reflex.

Some researchers claim there is a reduced electrical skin resistance at small discrete places they call acupuncture points. For several years I have diligently tried to confirm this observation in both patients and cadavers. I found there are thousands of smaller or larger skin areas of reduced resistance, some of which might correspond to acupuncture points, while most did not.

A doctor who knows acupuncture will be able to find acupuncture points electrically, by passing the searching electrode a few times over the desired acupuncture point. Each time an active electrode is passed over the skin its electrical resistance is reduced, and if this is repeated a few times one creates, de novo, one's own electrical acupuncture point.

According to my neurophysiological theory one would not expect to find discrete acupuncture points by electrical or other means. It is possible, though, that larger areas, related to the distribution of groups of neurones, may be found while in an abnormal state.

MERIDIANS

In some places the course of meridians follows the pathways of nerves or the position of dermatomes, in others it does not. I have shown in chapter I that in most (but by no means all) instances a neurological explanation fits in with more of the observed facts than with the hypothetical meridians.

Sometimes a needle in the leg produces a sensation (not a lightning pain) along the stomach meridian where it goes over the abdomen and thorax. This does not fit in with the route taken by a nerve trunk. The connections within the spinal cord are so numerous that further research might elucidate this and similar problems.

The experiments in chapter I have shown that most of the reflexes involved in acupuncture are spinal. It is possible that some reflexes, especially those which are not instantaneous, might involve higher centres. The experiments of Downman and Koblank illustrate that more than one neural path is excited by a single stimulus. Quite possibly a single pin prick in the leg may invoke two or three separate intraspinal pathways, and also a path along the sympathetic chain. Both the intraspinal and sympathetic routes have quick responses on the target area via a spinal reflex, and possibly delayed responses via suprasegmental pathways, which secondarily cause the release of hormones or vascular and other phenomena. I maintain, however, that the primary transmission system used in acupuncture is neural.

The fifty-nine or more meridians,* described by the Chinese seem to link interdependent areas of the body, even though they may be at a considerable distance from one another. For example: a mildly stiff neck may be helped by placing a needle in a gall-bladder acupuncture point on the foot, because the gall-bladder meridian goes through the neck. The nervous pathway between the foot and the neck is not obvious. The ancient Chinese, however, linked them together by a meridian, for they found the connection by experience.

The practical conclusion to be drawn from this is that, although the meridians do not exist as such, they illustrate in an almost abstract manner, the presumed neural pathways, which are as yet unknown. In that way the meridians are of paramount importance to the clinician whose main concern is to get his patients better. The meridians of acupuncture might even be compared to the meridians of geography: imaginary but useful. I hope that the investigations of neurophysiologists and others will map out the true neural pathways involved, which would then only partially correspond to meridians.

CATEGORIES OF ACUPUNCTURE POINTS

In chapter XI I showed that tonification and sedation, although they form a major part of traditional theory, are a philosophical conception which does not apply to the actual practice of acupuncture. If on pulse diagnosis the liver pulse is wide and hard, it is called overactive and requires sedation. If the point of sedation or tonification, or any other point whatsoever on that meridian or related points on other meridians, is stimulated, the pulse on Chinese pulse diagnosis becomes normal or nearly normal. Likewise if the pulse of the liver is narrow and soft, one could call it underactive,

*See my *The Meridians of Acupuncture.*

requiring tonification. The result would.be exactly the same if any of the before-mentioned points were used.

The obvious conclusion is that the twenty or so categories of acupuncture points (some books describe more categories) are superfluous. There is no physiological connection between the metal point on the heart meridian and the metal point on the liver meridian.

The different acupuncture points on the same meridian exert partly similar effects. Hence the tradition of joining them together with a line called a meridian. They have also partly dissimilar effects, for to some extent different neurones are stimulated.

LAWS OF ACUPUNCTURE

The various laws of acupuncture mentioned in chapters V, VI, and VII fall in many instances into disarray once one has discovered that tonification and sedation do not take place. Yin and Yang, coupled organs, full and empty diseases, cold and hot diseases, become untenable.

The law of the five elements demonstrates some connections between organs well known to physiology, and some connections which presumably exist but are not yet known. In clinical practice one finds that certain of the connections happen frequently, while the others occur rarely or never. The frequently occurring connections can in most instances be more easily explained in Western terms than via the Chinese pentagram.

If all the laws of acupuncture are taken together it will be seen that every phenomenon occurring in health and disease can be explained, and hence it leaves one with little explanation at all.

CONCLUSION

What I have written in this short chapter, and also in chapters I and XI, might give the reader the impression, that there is little left to acupuncture, for I have demolished practically the whole of the traditional theoretical framework. This is far from being the case, for I practise acupuncture exclusively about 90% of my time, and I would not do so if I did not achieve *better* results than in practising Western medicine in the appropriate type of disease or dysfunction. There are, of course, many diseases where Western medicine is better than acupuncture.

The crux of the matter is that I try to combine acupuncture with the principles of Western physiology, anatomy and medicine in general. In some instances Chinese theory explains phenomena better than Western theory, and hence I treat the patient accordingly. I try to keep my feet in both camps, for there is much that is unknown to Western physiology that can in a certain way be explained by Chinese tradition, or at least enough of it can be explained to know how to treat a patient from an acupuncture point of view.

A traditional Chinese doctor practising acupuncture will achieve good results, albeit for the wrong reasons, which is of little concern to the patient. A traditional doctor may say that a specific very small acupuncture point should be stimulated; a Western doctor may say that anywhere within a given area is sufficient. Both doctors achieve equal results if the Oriental's acupuncture point lies within the area of hypersensitivity of the Westerner. Again: a Chinese doctor may say that the fire point on the Yin wood meridian should be used; his Occidental counterpart may suggest that any 'point' on the liver meridian below the knee is sufficient. The two colleagues will have equal results, for the fire point is within the Western grouping. There are many more instances where the Oriental and Occidental doctors will have the same cure rate, largely because the traditional theory has made acupuncture more complicated than it actually is. I hope Western science will be able to transform acupuncture to the same extent as it has that other great Chinese invention, the magnetic compass.

Despite what I have written I continually read traditional Chinese books on acupuncture. Apart from the theory, which may be wrong in our eyes, though sometimes right as an abstraction, the Chinese books contain a wealth of clinical details and practical observations, which are extremely useful. It is for this same reason that I would advise doctors who wish to learn acupuncture, or others who are interested in the subject, to read those parts of my books that are purely traditional.

THE ENERGY OF LIFE—QI

The Chinese theories related to Qi (the energy of life), Nourishing Qi, Protecting Qi, Blood, Life Essence, Spirit, Fluid and similar connections mentioned in chapter IV, are most easily understood as a traditional Chinese concept, linked to a view of the world different to that of most Occidentals. Western doctors who practise acupuncture, or neurophysiologists who investigate its mode of action, can do without this traditional idea.

If a patient, even a Western doctor, has been ill and then recovers, he will say "I feel better, I have more energy." If this same doctor is then asked what is energy (called in Chinese Qi), he will probably say that such a thing does not exist. A contradiction and at the same time not a contradiction.

From the point of view of Western medicine, disease ensues when the biochemical processes of the body are disturbed. If, for example, there is a deficiency of potassium, the body chemistry is altered and the patient has, among other symptoms, little energy. The energy cannot be measured directly; only its secondary effect in reducing muscular activity may be measured.

The Oriental doctor considers energy as something primary and 'real,' whose deficiency secondarily causes disease. The Occidental doctor thinks the chemistry of the body, which only secondarily affects energy, is primary. Textbooks of physiology do not mention the conception of biological energy as something primary.

These two points of view are only partially contradictory. They are only looking at life from different points of view.

Much of Chinese medical theory describes what the patient feels. The patient feels differences in energy. He often feels something along the course of meridians. The Western doctor often excludes the patient's feelings and measures the serum electrolytes, haemoglobin and faecal fat instead.

Few people would disagree that when they see a meadow it is green. A physicist would say, however, that the meadow emits a certain wavelength of light which is then *subjectively* interpreted by the eye and brain as the colour green. This is little different from the person who is hit with a sledge hammer and then subjectively interprets it as the taste of onions—something a dog could probably be trained in the Pavlov manner to do.

If the physicist is asked what a wavelength of light is, he might explain it in more detail, using Einstein's particle theory of photons, which is not considered to be a physical reality. From which it emerges that Chinese metaphysics is hardly less real or unreal than the theoretical background of modern physics, which is the foundation of most modern medicine.

12-72

BIBLIOGRAPHY

In Chinese

Zhenjiuxue Jiangyi (Lectures in Acupuncture and Moxibustion); compiled by the Acupuncture Research Section of the Shanghai Academy of Chinese Medicine; published by the Shanghai Scientific and Technical Publishing House, Shanghai, 1960.

Zhongyixue Gailun (A Summary of Chinese Medicine); compiled by the Nanking Academy of Chinese Medicine; published by the People's Hygiene Publishing House, Peking, 1959.

Zhenjiuxue (The Study of Acupuncture); compiled by the Acupuncture Research Section of the Nanking Academy of Chinese Medicine; published by the Jiangsu People's Publishing House, Nanking, 1959.

Changjian Jibing Zhenjiu Zhiliao Bianlan (A General Survey of Common Diseases and their Treatment by Acupuncture in Tabular Form); compiled by the Peking School of Chinese Medicine; published by the People's Hygiene Publishing House, Peking, 1960.

Jingluoxue Tushuo (An Illustrated Survey of Meridians); compiled by Hiu-jan and Zhu Ru-gong assisted by Wu Shao-de, Wu Guo-zhang and Zhang Shi-yi; published by the Shanghai Scientific and Technical Publishing House, Shanghai, 1959.

Yuxuexue Gailun (A General Survey of Acupuncture Points); compiled by Hiu-jan and Zhu Ru-gong, assisted by Wu Shao-de and Zhang Shi-yi; published by the Shanghai Scientific and Technical Publishing House, Shanghai, 1961.

Zhengiufa Huilun (General Thesis of Acupuncture Treatment); compiled by Lu Hiu-jan; published by the Shanghai Scientific and Technical Publishing House, Shanghai, 1962.

Zhengiu Yhyetupu (Description of Acupuncture Points in Tabular Form); compiled by Lu Hiu-jan; published by the Hong Kong Hong Yci Publishing Co., Hong Kong, 1967.

Zhengiu Dacheng (Summary of Famous Ancient Works on Acupuncture— Ming); by Jang Gi-zhou; published by The People's Hygiene Publishing House, Peking, 1963.

Zhengiuhye Shouce (Handbook of Acupuncture); compiled by Wang Hye-tai; published by The People's Hygiene Publishing House, Peking, 1962.

Huangdi Neiging Suwen Gizhu (The Yellow Emperors Classic of Medicine in Dialogue Form with Annotations—Ching); by Zhang Jin-an; published by The Shanghai Scientific and Technical Publishing House, Shanghai, 1963.

Zhengiu Giajibing (A Classic Thesis on Acupuncture—Gin); by Huang Pu-mi; published by The People's Hygiene Publishing House, Peking, 1964.

Zhengiu Gefu Hyangie (Selection of Songs and Rhymes on Acupuncture with Explanations); by Chen Bi-liu and Zhen Zhuo-ren; published by The People's Hygiene Publishing House, Peking, 1962.

Zhungguo Jihye Dacidian (The Encyclopaedia of Chinese Medical Science in 4 volumes); edited by Shai Kwan.

In Other Languages

L'Acupuncture Chinoise; by Georges Soulié de Morant; published by Jacques Lafitte, Paris, 1957.

Essai sur l'Acupuncture Chinoise Pratique; by J. E. H. Niboyet; published by Dominique Wapler, Paris, 1951.

Complements d'Acupuncture; by J. E. H. Niboyet; published by Dominique Wapler, Paris, 1955.

Traité de Médicine Chinoise — several volumes; by A. Chamfrault; published by Coquemard, Angoulême, 1954 and later.

La Voie Rationalle de la Médicine Chinoise; by Jean Choain; published by Editions, S.L.E.L. Lille, 1957.

L'Acupuncture Chinoise; by P. Ferreyrolles; published by Editions S.L.E.L., Lille, 1953.

L'Acupuncture; by Yosio Manaka, Odawara, Japan, 1960.

Acupuncture; by Henri Goux; published by Maloine, Paris, 1955.

Acupuncture; by H. Voisin; published by Maloine, Paris, 1959.

Einführung in die Akupunktur; by Johannes Bischko; published by Haug, Heidelberg, 1970.

Die Akupunktur eine Ordnungstherapie; by Gerhard Bachmann; published by Haug, Ulm, 1959.

Die Chinesisehe Medizin; by Fr. Hübotter; published by Asia Major, Leipzig, 1929.

The Yellow Emperors Classic of Internal Medicine; by Ilza Veith—translated from classic Chinese; published by Williams and Wilkins, Baltimore, 1949.

DR. FELIX MANN, a London physician, was trained at Cambridge University and Westminster Hospital. Becoming dissatisfied with the results of orthodox medical treatment in his own practice, he spent several years working as a doctor in other countries and studying new medical theories. While in France, Dr. Mann witnessed the curing of an appendicitis patient by a needle-prick below the knee, and decided to turn to the study of acupuncture himself. He worked extensively with European physicians and later studied with a Vietnamese doctor. He also learned Chinese so he could study the acupuncture texts in their original language.

Back in London, he reopened an office and began to apply what he had learned. He had much success in the practice of acupuncture and generated so much interest in the subject that he founded The Medical Acupuncture Society, an organization specifically designed to further the practice, teaching and research of this unique and ancient form of medicine.

VINTAGE WORKS OF SCIENCE
AND PSYCHOLOGY